KEEP
CALM
AND TRUST
THE
SCIENCE

Professor Luke O'Neill is a world-renowned immunologist and a professor of biochemistry at Trinity College Dublin. He is the author of three bestselling books, the award-winning *Never Mind the B#ll*cks, Here's the Science*; *Humanology*; and *The Great Irish Science Book* for children. Luke is a member of the Royal Irish Academy and is a fellow of the Royal Society.

KEEP
CALM
AND TRUST
THE
SCIENCE

AN EXTRAORDINARY YEAR IN
THE LIFE OF AN IMMUNOLOGIST

PROFESSOR
LUKE O'NEILL

Gill Books

Gill Books
Hume Avenue
Park West
Dublin 12

www.gillbooks.ie

Gill Books is an imprint of M.H. Gill and Co.

© Luke O'Neill 2021
9780717191819

Designed by Bartek Janczak
Print origination by O'K Graphic Design, Dublin
Edited by Djinn von Noorden
Proofread by Neil Burkey
Printed by CPI Group (UK) Ltd, Croydon, CR0 4YY

A CIP catalogue record for this book is available from the
British Library.
5 4 3 2 1

To all the scientists working on COVID-19,
whose work will finally release us
from the pandemic.

INTRODUCTION

At the beginning of 2020 things were looking really good for me. I'd been a research scientist from 1985, when I did a research project on Crohn's disease. I then trained as a scientist in the UK, continuing to work on inflammatory diseases, moving into rheumatoid arthritis. Finally, I established my own lab in Trinity College Dublin and, with my team, made some interesting discoveries about the immune system, and how it goes wrong in various diseases. I'd published lots of papers, and even won some awards, including becoming a Fellow of the Royal Society in the UK. Having 'FRS' after your name makes you something of a Jedi knight, although as one friend told me at the time, it stands for 'Former Research Scientist'.

So the science was going well, and that had allowed me to do two other things that give me great satisfaction and pleasure: communicating science and turning scientific discovery into new treatments for patients.

I became an academic because I like teaching. I'd been doing that for the general public too. I'd had a weekly slot with Pat Kenny on Newstalk radio for a few years. I had written two columns for the *Sunday Independent*, with a promise of more to come. I had done a bit of TV, including making a documentary on RTÉ about the Austrian physicist Erwin Schrödinger. He had worked in Dublin and given his famous 'What Is Life?' lectures in 1943, which sparked the revolution in biology that led to the structure of DNA being solved, explaining how genetics works. I'd loved all that.

In December 2019 I'd been asked to help with a *Prime Time* item on synthetic meat and had almost finished writing my second science book for a lay audience – named, to my great joy – *Never Mind the B#ll*cks, Here's the Science*. I'd already published *Humanology*, about the science of being human, and a science book for children called *The Great Irish Science Book*, both of which had done well. I'd grown to love all of it – the radio, a bit of TV and the books – because I want to communicate science to as many as I can reach.

In early 2020 we were also making advances with new medicines, begun in my lab, for patients with serious diseases such as Alzheimer's and Parkinson's. The company I had co-founded, Inflazome, was attracting interest from a couple of big pharmaceuticals and I thought it possible that one might offer to buy us. This was an especial thrill. Even though I had published lots of papers, I wanted my discoveries to ultimately help patients, and that was becoming an increasing possibility. I'd failed in that before with a company called Opsona, so there were no guarantees; but it was looking more and more likely.

All of this is what I thought I would be doing with my science in 2020: generating data, benefiting patients – and hopefully having fun along the way. It was going to be a special year.

I was also becoming aware of a new coronavirus. And so I decided to keep a diary, something I hadn't done since I was a teenager. Every night, usually well after midnight, like a 21st-century Samuel Pepys, I would record what had happened during the day. When everyone else was in bed asleep, I'd be working.

Yet I never imagined that, like Pepys, I'd be writing about a plague.

JANUARY 2020

THURSDAY 16 JANUARY

Woke up in the Pickwick Hotel, San Francisco. Saw the headline: 'Chinese respiratory illness claims first life'. So there's this new virus in China. Intriguing, but nothing to worry about. It could be like SARS.

Went for a delicious steak dinner with Jeremy, Angus and Thomas, my Inflazome colleagues. Too much red wine, but we deserved it. Spent the last four days trying to interest all the big pharmaceutical companies in us. We have a drug that might treat Parkinson's, Alzheimer's, ulcerative colitis, asthma, heart disease, you name it. And it's not snake oil. It's an NLRP3 inhibitor, so there. NLRP3 is a really important inflammatory protein that goes wrong in so many diseases, and we might have found a great way to stop it.

Damp and cold outside. Loads of meetings. All the drug companies here as usual and we made our pitches again. Several are interested in us, which is great. Good to have competition. But Roche seem especially keen. They sent 20 people to meet us. We sat on one side of the table, and all 20 sat on the other side in a huge big line, poker face after poker face.

The latest on this Wuhan virus, though, is that it has killed someone. The Chinese don't seem too worried because they think it just passes from animals to humans. No evidence yet for human-to-human transmission. But they are watching things closely.

I avoid virologists at conferences as they always bang on about the risk of a global pandemic. Fun-crushers! Looked into it a bit. On 31 December the Wuhan Municipal Health Commission put information on its website saying there was a pneumonia outbreak in the city of 'unknown aetiology'. Reuters picked up on it. The origin was likely to be Huanan seafood wholesale market: it had been heavily disinfected and stallholders were all told to wear masks. Hong Kong responded by saying they would put anyone coming from Wuhan into a 14-day quarantine. They had SARS before and don't want a repeat.

The same day the Shanghai Centre for Disease Control said they were able to contain it and no human-to-human transmission had been reported. That's enough of that.

Today I have to fact check more chapters in my new book, *Never Mind the B#ll*cks, Here's the Science*. So pleased with myself that I came up with the title, which came to me on a flaming pie over Christmas.

Right, up and at 'em! First a big American breakfast in the Pickwick's restaurant. It has lots of drawings of Dickens's *Pickwick Papers* on the wall. I like this hotel: old fashioned and comfortable and my desk at the window has a good view of downtown San Francisco. All the cars rushing by, each with people with their own separate cares.

I'll do the chapter on vaccines today.

SATURDAY 18 JANUARY

About to take off. Long haul to London and then Dublin. It's been a good week. We met them all – GSK, Pfizer, Eli Lilly, Takeda, BMS, Novo Nordisk, Sanofi. All the big boys. Literally boys. I noted how many women were in the meetings – 10 per cent, I'd say. Why is that? I much prefer meetings with a mix. All the men look and dress like me. Jacket and trousers. Shirt but no tie. Checking our phones every five minutes.

Heard from Eithne of *Prime Time* that the clip I sent of me eating an Impossible Burger in Burger King in San

Francisco, for her piece on the future of meat, worked well. She had interviewed me on 3 January in my lab for a piece she's doing on lab-made meat and then asked me to eat one in Burger King. Me and Jeremy had fun filming it – a welcome break from all the meetings. It will be the first time on *Prime Time* for me – delighted Eithne asked me and it was great working with her. I suspect this will be my one and only time on *Prime Time*, so I thought: why not?

Settling into my nine-hour flight, but it's OK – no one can get at me! I pull the blanket over me as the US and the Atlantic Ocean rush beneath me. A sense that I'm going somewhere. Reflected a bit on Inflazome. It began with the discovery in my lab in Trinity of a drug that blocked the inflammatory protein NLRP3. We knew it might be useful for many diseases. That led to a conversation at a conference in Australia with Matt Cooper, a chemist working in the University of Queensland. Between us we hatched the plan, with backing from Manus Rogan of Fountain Healthcare Partners, who invests in new companies. Matt's lab improved on the initial drug (which had been found by Pfizer) and we tested what he made. Now we have some very interesting ones that we think could really work. We need a big pharmaceutical company to take them on now, and get them to patients. Imagine that.

What a dream that would be. And it might well happen this year.

I thought about other academics who make discoveries and then try and make them count by forming companies. Many don't realise us ivory-tower types can raise finance and get companies going and yet it's not uncommon. Here's hoping it all works out.

MONDAY 20 JANUARY

In Rotterdam. Oh, this is a good one! Conference on viruses and immunometabolism has begun. All about how viruses change how immune cells use nutrients. A new idea really, as there is evidence that when a virus infects a cell, it changes how that cell uses glucose and fats. If we can understand more about that we might come up with better ways to stop viruses. There was some chat in the coffee break today about this new Wuhan virus in between the usual topics us scientists talk about. How we were shafted by a journal that wouldn't publish our work, or a grant agency that won't give us money. Or some rival dissing our work. It's therapy.

Some of the scientists here are experts on SARS so they are hungry for more information. They don't seem especially worried. One said to me that it will be quickly contained as the Chinese learned from SARS and can efficiently isolate infected people. We did learn, though, that 41 people are reported to have died so far. A funny-looking pneumonia, it seems.

Great meeting John Hiscott again – old friend and collaborator. He's currently working in Rome doing more work on interferons and how they limit viruses. He has new stuff on Dengue virus. John typifies what I love about this job. Very generous guy. Always encouraging. I remember when I first got to know him at a conference I organised in Dublin in 2003. Five hundred immunologists in Dublin Castle for the big reception. I was at the top of the stairs welcoming them all and he came up to me and shook my hand and thanked me for getting him to Ireland, where he had the inevitable ancestors who had emigrated to Canada in the 1800s. He gazed around St Patrick's Hall in awe, and I could see a tear in his eye. The Old Country. I said to him, John, it was the Brits who built this place. He laughed out loud. Can't beat two colonials slagging off the Brits.

WEDNESDAY 22 JANUARY

The podcast I did with Blindboy a couple of weeks ago has gone down well. It was a great romp through immunology. I'm a big fan of Blindboy – he has huge warmth and humanity about him.

After dinner with the virologists, I walked back

through the docks in Rotterdam – cold, foggy and atmospheric. Made me think of what the city must have been like in the old days – a bustling port, taverns full of sailors. I feel like I'm in a movie when suddenly there's a timeslip and I'm dressed in 17th-century clothes and about to get on a ship. Weird. Must have been the copious wine at dinner.

The virologists are anxious. One from New York said there was a report of the first case of this new virus in the USA, so it hasn't been contained by the Chinese. And the evidence now indicates that there *is* human-to-human transmission. It suddenly looks more serious. He said that if it hits the US it might mean more research funding, which has been neglected in the area of coronaviruses. Typical scientist – always looking for funding! He is still hopeful that if it's related to SARS it might be possible to quickly contain it.

Went to bed with the song 'Rotterdam (Or Anywhere)' by the Beautiful South playing on Spotify (ah the joy of Spotify!). Nice little guitar riff on that one.

THURSDAY 23 JANUARY

On with Pat Kenny on Newstalk. Did it from my hotel in Rotterdam. Talked a bit about the Wuhan virus. In my mind because of all these virologists. Gave the history, and Pat had some killer questions – will it be just like SARS? Told him SARS had emerged in 2002, got to 37 countries, infected 8,000 and killed 69. MERS then cropped up in 2012, infected 2,500 and killed over 800. All planes and trains have been stopped coming out of Wuhan and 41 events have been cancelled. Pat and I agreed that this was one to watch.

Thinking about it afterwards, it does look like China might be suppressing information. The BBC says that scientists and doctors were told not to publish anything and to transfer all samples to one institution. An ophthalmologist has reportedly been officially reprimanded for making false comments on the cases of the SARS-like disease in the Huanan market. Another scientist did, however, release the sequence for the virus on 10 January. Secretly sent it to a lab in Melbourne where it was passed to one in Edinburgh. A new virus, alright.

Also looks like that by early January, Chinese scientists had ruled out 26 other respiratory viruses and by 3 January they had found a new virus, naming it 2019-nCoV. Imagine being the person to see that virus first. Mind you, may not be that big a deal if it's like SARS, which could be contained. The first death was recorded on 9 January – a 61-year-old man with chronic liver disease. The WHO have said that China has responded quickly to contain the virus. Good news.

Also read today that China has reported finding a bat virus with 96 per cent similarity to the new coronavirus – so it might have come from a bat, just like SARS. A publication has also stated that the entry point for the virus into cells is the same as SARS: a protein on lung cells called ACE2. So there's no doubt in my mind that this new virus is highly related to the SARS virus.

FRIDAY 24 JANUARY

Saw something worrying today in *The Lancet*. People can be asymptomatic for several days before developing the disease, which might increase the risk of contagiousness. This is not like SARS, where people with symptoms are the ones who spread the disease. *The Lancet* also published a very good paper confirming human-to-human transmission. We could be in much bigger trouble than we thought.

SUNDAY 26 JANUARY

On a flight to Denver for the big Keystone immunology meeting in Boulder. I had trouble getting to sleep last night. I was thinking about the piece I'd written for the *Sunday Independent* on this new virus in China. Name is currently 2019-nCoV, or 2019 new coronavirus.

Don't want to frighten people – there's a lot we don't know. So I said there was nothing to be too frightened of *yet.* It has no name yet – I suggested 'Wuhan Respiratory Syndrome' or WURS. That would make it easy to remember as it's like its relatives, MERS or SARS. I wrote that if it's a bad descendant of SARS, it might cause worse disease, although so far that doesn't seem to be the case. I did say that one worry is that they now know that it can spread from human to human. A man working at the seafood market in Wuhan caught it and then spread it to his wife who hadn't been to the market. It causes pneumonia. Those who become sick have a cough, fever and breathing difficulties. Those who have died are known to have been already in poor health. The Chinese are trying to stop it spreading. Travel restrictions have been imposed. The Chinese and the WHO are hopeful that it can be contained like SARS, but they have highlighted the role of 'super-spreaders' – one person spreading it to many, as seems to have happened in Wuhan. The WHO have also said that the world now needs to act as one against this new virus.

Every five minutes I'm getting another update. Since writing the piece, disturbingly, the whole province of Hubei has gone into quarantine. That's millions of people in quarantine. They must be worried. And Hong Kong has declared a state of emergency, closing all schools. Jeez.

TUESDAY 28 JANUARY

Went to some interesting talks at the conference, especially one on coeliac disease by my old friend Ludvig Sollid – whose surname I always thought reflected his science … Ludvig the Solid. Got to know him on an immunology committee I chaired at the European Research Council for three years. He's a world expert on coeliac disease and his talk was a great update on this issue, which is common in Ireland, most likely for genetic reasons.

The deadline for *Never Mind* looms – finished the addiction chapter in my hotel room, and then gave a talk to undergraduate students from the University of Colorado. They would not normally attend the main conference but the organisers are keen to involve them. Thirty of them traipsed into quite a small room so they were a bit on top of me. It was a great atmosphere and a great session. They were keen and asked lots of questions on cytokines and innate immunity. The next generation! No mention of the Wuhan virus so it hasn't quite registered yet, which is good.

Word has come from Germany indicating that indeed, this virus can spread from one person to another with no symptoms. This worries me. It looks like it's spreading widely. Only a matter of time until it hits Ireland. Also – scientists in Australia have reported that they can successfully grow and study the new virus. All we can do right now is hope that public health measures can control it. I feel unease, however.

WEDNESDAY 29 JANUARY

Gave my talk at the conference. A tiny bit nerve-racking. Wi-Fi was down in the hotel so I told the audience that for once they should pay full attention as opposed to being on their phones (which is the way of things these days). Talk went well, though. Gave them our new data on itaconate as a possible anti-inflammatory agent. Itaconate is made in inflammatory cells; our data points to it being a natural anti-inflammatory factor. May well have potential for several diseases, just like our NLRP3 inhibitors.

Taxi ride all the way from Boulder to Denver Airport with three other scientists. I always feel a bit melancholic heading home after a great conference. You get used to the rhythms of each day and I enjoy hanging out with other like-minded scientists. The buzz of the chat. People are always surprised when I tell them how sociable scientists are. We're like everyone else, really. The taxi drove through beautiful scenery; the snowy peaks of Colorado. Saying goodbye to my fellow scientists, we all shared how much we were looking forward to our next conference in the coming months.

Denver airport is magnificent. Huge white structures that look like snowy mountaintops and a huge sculpture of a horse on its hind legs. Saw in the airport that Air Canada, British Airways and Lufthansa have cancelled all flights to and from China. Air Canada says until 29 February. They must think the whole thing might be in abeyance by that date.

THURSDAY 30 JANUARY

Landed in Heathrow and got another flight to Manchester, where I'm assessing the Medical School in the University of Manchester. Dermot Kelleher, a former colleague from Trinity now in Vancouver, and Alan Irvine from the Children's Hospital in Crumlin are also part of the committee. We gave a good grilling and then the three of us had a few drinks in the bar. Three Paddies together again. I defied the jet lag (not much sleep in 48 hours) with good whiskey and excellent conversation. Dermot asked me what I thought of this new virus. I said it's interesting alright but we both agreed nothing to worry about yet.

The WHO, however, declared today that the virus was a 'Public health emergency of international concern' and that all countries should get ready for testing and isolation of those infected. The fear level went up a notch reading that. I wonder how many countries will heed that call? It's now very certain that there is human-to-human transmission going on. This was uncertain earlier this month but I wonder if the Chinese knew all along and didn't act on that?

China is already developing a vaccine. Now that's fast! Could take months. Let's hope it's not years. The quickest vaccine ever – for mumps – took four years. Good Lord. If it takes four years and this gets really bad, then we're fucked. Steady now, Luke. Steady.

FRIDAY 31 JANUARY

Home at last! Pure exhaustion. Slept the sleep of a dead man for 20 hours. You know, that feeling when you lie down on clean sheets and immediately relax fully and fall into a deep sleep. Great to see my wife Marg and younger son Sam again. Absence makes the heart grow fonder … but I must avoid familiarity breeding contempt. It's funny how sayings always have an opposite. Maybe we should just combine them. Absence breeds contempt? Many hands spoil the broth?

I wake up and read on my phone that a new committee has been set up by the government to co-ordinate the national response to the virus. Catchy name, though: National Public Health Emergency Team (NPHET), pronounced 'Nefet'. Looked up the membership and can't see a single scientist on it. Bah!

Italy has now declared a state of emergency. The first EU country to do so. So has the US, who say they are closing their borders to all foreign nationals. And Trump just banned all flights from China.

A lot of talk about bats in the science literature today. They seem to tolerate coronaviruses by ramping down their inflammatory response. This might involve my favourite protein, NLRP3, which can drive inflammation during infection but is absent in bats. Is it possible that we humans invaded the bats' territory and so the virus then infected us? Revenge for us ruining their environment? This debate will rage on.

Had a cup of tea in front of the telly and fell into another deep sleep. I like it when January ends. Yet looking back, it was a good one. A successful trip to San Francisco, which might help us sell Inflazome. Progress with *Never Mind*. Two great conferences, and the hope of publishing our work.

Typical enough, I guess, except for one thing. This virus. Can't help but think of the movie I watched on one of those many flights. It was called *The Gathering Storm*.

FEBRUARY 2020

SUNDAY 2 FEBRUARY

Another month begins. St Bridget's Day. I always feel winter is starting to yield when we get to February. I think I can see the tips of buds on the trees just outside the front door getting a bit sticky. Weather still red raw though.

Saw in the news that several landmark buildings in the United Arab Emirates were illuminated on Sunday night to show support for Wuhan and Chinese communities around the world. This was to recognise the suffering they've been going through. Nice gesture.

WEDNESDAY 5 FEBRUARY

A lot of teaching this week. Lectures to final-year

immunologists and biochemists. I give my usual cytokines lectures. I really enjoyed giving these – you can't beat cytokines! These are the proteins that control every aspect of the immune response in our bodies. I've been working on them since 1985, when not many were known. There are now hundreds of them. I tell the students how blocking them can really work in some inflammatory diseases. Such a complex business. All the acronyms must drive the students mad: JAK, STAT, SOCS, NF-kappa B, MYD88, MAL (the one I named, yay!), TRAM, TRIF. Like a whole new language. Always mention how knowing this stuff might help us get new therapies for diseases like rheumatoid arthritis. It's striking how many new medicines have come from the world of cytokines.

Met with Trinity's technology transfer office in the Science Gallery café to talk about Inflazome. Gave them an update on how things had gone in San Francisco and how you never know, we might be bought and Trinity, as a shareholder in the company, might make some money! Very long shot – but you never know. I could tell by the way they looked at me that they could sense something in what I told them.

THURSDAY 6 FEBRUARY

Today's Pat Kenny show was in front of a live audience, broadcast from the Cliff House restaurant on St

Stephen's Green. A sponsor, I guess. I discussed the latest science on why we go grey. It turns out that cells in our scalp makes a natural type of bleach, which makes our hair go grey. Most of the audience were grey, like me and Pat, so I told them all we're all in the same grey-headed club and they laughed.

Didn't talk about coronavirus even though the US reported its first death today – a woman died of myocarditis (heart inflammation) caused by the coronavirus. The WHO have also said that there are no known therapeutics or drugs that are effective against the virus. This is to counter a lot of reports on possible therapies working, with the WHO saying they don't work. It seems all kinds of snake-oil salesmen are coming out of the woodwork. I suspect there will be a lot more of this kind of thing.

Went for my usual slot on *The Six O'Clock Show* with Muireann and Martin. (I've been on a few times. The first time I was on I plugged my previous book *Humanology* when, during the interview on the science of attraction, I asked Muireann if she was ovulating. Deirdre O'Kane, who was also a guest that time, found this very funny. Later we were in the kitchen, standing around chef Kevin Dundon as he poured hot chocolate sauce over a sponge cake. Deirdre shouted, 'Jaysus, I've just ovulated!')

We talked about how optimism is good for your immune system, and how to stay optimistic in these wintry days and we always have a laugh or two. Badly needed in these dark days of February! After the show I got a taxi into the city centre, where the science students in Trinity asked me to be the MC for a fundraiser quiz in J.W. Sweetman's pub. It was packed to the rafters, everyone shouting and roaring. Myself and two colleagues, Emma Creagh and Áine Kelly, shared out the questions. Then they handed me a guitar to play songs for the music round. What am I, a performing seal? Didn't mind at all, of course. Great to see the students doing this kind of thing, kicking back a bit and enjoying themselves.

SATURDAY 8 FEBRUARY

Guest of honour at the Irish Science Teachers' Association annual dinner. I am the incoming president. The outgoing president is weatherman Ger Fleming. He was supposed to come along and hand the chain of office over to me, but two things happened on the way to the dinner. First, he was called away to some weather emergency in Eastern Europe; and then, while he was away, the ceremonial chain was stolen from his house. An inauspicious start to my term of office.

I made an after-dinner speech and told them the story of how I had met up again with my old biology teacher from Bray. I'd been on the radio last summer, interviewed by Keelin Shanley. It was a great interview as I was very fond of Keelin. She had a degree in biochemistry from Trinity, so we chatted about the science of what I'm doing in the lab. She asked me how I'd got into biology, and I mentioned my old teacher Fran Mooney. He'd really inspired us all – in fact, I can still remember the day he told us about DNA.

A week or so later I got a letter from him to thank me for mentioning him. He said he might turn up at my next public lecture. And lo and behold, he did. It was in the Smock Alley Theatre. He came up to me at the end and I put my hand out. He grabbed me and gave me a big hug. As we separated he said, 'Er, I don't remember you.' Thanks, Fran. We're blessed in Ireland with our cohort of science teachers who make a huge difference. It's one reason why Ireland does well when it comes to overall scientific literacy. Long may it continue and to hell with the begrudgers!

TUESDAY 11 FEBRUARY

Dinner for the board of Inflazome after the first board meeting of the year. I love meeting the investors and board members who are, more often than not, scientists themselves. They come to Dublin from Europe and the US and it's great to chat to them, not just about Inflazome, but other things they are working on. They are very smart people – often have medical degrees, PhDs, MBAs. I guess they have to, as they are handling so much money! We had a delicious dinner in the Marker Hotel. Lots of good wine. This is standard – a reward for all the hard work, but also a chance to relax and talk issues through in a more convivial setting, which can help the business side no end. Ended up in the bar after as Dhaval, who is a key board member, formerly of Novartis, loves Irish whiskey. Great chat about possible future companies.

News on this coronavirus continues to emerge from China, and we spoke a bit about that. Dhaval said it looks a bit more serious with Wuhan going into quarantine. But we all agreed that hopefully, like SARS, it will be contained. No sign from what I can tell that it's any worse.

And the virus has a name! The WHO have named it SARS-CoV2 (severe acute respiratory syndrome coronavirus-2). Because it's a close relative of SARS, which has been renamed SARS-CoV1. It must have been like that after World War Two, when the Great War was renamed World War One. And the disease it causes has been named Coronavirus Disease-19 or COVID-19, with

19 because that's the year it started. I'll bet people get that mixed up – they'll think its virus number 19. Still, at least we have a name for the virus and disease now.

WEDNESDAY 12 FEBRUARY

Went to Keelin Shanley's funeral today. Huge crowd. She was so well loved. Strange I remembered her the other day at the science teachers' dinner. Had a brief chat with her dad, Derry – knew him from Trinity when he was Dean of the Dental School. He said how I had given Keelin comfort, which I wasn't aware of. I'd met her a few months back for a drink in Fitzgerald's when she gave me an update on the experimental treatment she'd been getting at the NIH for her breast cancer. I'd told her how I had lost my own mother at 17 to breast cancer. She said it gave her hope for her own teenage children. Good Lord, it can be a tough old business, this life. I won't forget Keelin.

THURSDAY 13 FEBRUARY

Pat and I had our first big chat about COVID-19. I explained the difference between SARS-CoV2 and COVID-19, and I said one was the name of the virus and the other the name of the disease, like HIV and AIDS. We talked about whether it can be contained in China.

A good day in the lab today: Z's paper got accepted for publication in *Nature Communications*. A very good journal. It was a long journey and I'm happy for Z as he showed such perseverance with it, dealing with lots of criticisms from editors and reviewers, but we won in the end. This happens. People don't realise how tough it can be on scientists. It's an important paper about a protein called caspase-4, which we implicated for the first time in asthma. This might stimulate drug companies to go after caspase-4. We raised a few glasses to celebrate, as per usual.

Outbreak on a cruise ship, the *Diamond Princess* currently in Japan. There have been 218 cases – what is called a high attack rate. It's worrying because it confirms that COVID-19 is highly contagious. An 80-year-old passenger got off the ship in Hong Kong on 25 January. He began to feel unwell and six days later he was admitted to hospital, where he tested positive for COVID-19. They then tested everyone on board and registered all those positives. The ship was put into quarantine in Yokohama on 4 February and it's still there.

SATURDAY 15 FEBRUARY

I submitted *Never Mind* to Gill today. I made lots of back-ups, because imagine if I or they somehow lost it? 80,000 words would have to be rewritten. Even put it on a memory stick for good measure. I read bits of it randomly and still liked it, so I guess that's what's important. Satisfaction.

I also gave my first ever online lecture to a big audience organised by Eleanor Fish, the eminent virologist. She said the idea is to allow as many people as possible to attend. Strange experience, talking into my computer screen with no audience reaction. Can't imagine this being the future. We're social creatures.

SUNDAY 16 FEBRUARY

On the *Brendan O'Connor Show* on RTÉ Radio One. I am a big fan of Brendan's. We have the same sense of humour, I think, and I remember his one and only hit single 'Who's in the House? Jesus in the House!' He'd heard me on Pat Kenny talking about COVID-19 and wanted a quick chat. We also spoke about going grey, as I had a short piece in the *Sunday Independent* on that. Brendan seemed more worried about that than COVID-19! Strange enough in that I was a big fan of Marian Finucane, who sadly passed away and who Brendan replaced. I'd also thought it would be great to be interviewed by her. It wasn't to be, and yet there I was on with Brendan.

MONDAY 17 FEBRUARY

Hosted a small meeting in Trinity of immunologists, who came over to discuss projects in the area of immunometabolism in my other company, Sitryx, which is developing new drugs that target metabolism in the immune system as another way to treat inflammatory diseases. We went through some of the projects. Such a productive day.

Over dinner though, all we talked about was COVID-19. I took them to the Ginger Man pub and asked them all, just how serious is this going to be? Jon Powell from Johns Hopkins said cases are going up exponentially in China and nearby countries. Doreen Cantrell from Dundee said the medics in her university are getting concerned as they know that if it was to come to the UK, the hospital system couldn't cope. Haven't seen anything quite like this, where someone says something, we are all silent and then someone else says, 'Can you repeat that?' A part of me doesn't believe any of it. Or perhaps what I'm actually feeling is, *I don't want that to be the case.* There was definitely a sense of foreboding in the air tonight. We couldn't change the topic of conversation.

FRIDAY 21 FEBRUARY

Well, a most unusual day. I went to my old hometown

of Bray today to give a talk on my *Great Irish Science Book*. Linda came down from Sligo for it. Great seeing her again – she did such fantastic drawings for it. We did a few demos for the kids. The book has gone down a storm: over 20,000 copies sold and great feedback. Someone came up to me to give out: they can't get their son to go to school in the morning because all he wants is to read the book!

Had some pints. I wondered would people ask about COVID-19 but no one did. Is it only me who's worried? Taxi to Dalkey for the St Patrick's Parish quiz night. The excellent host Gary German always gets a question in for which the answer is the Neil Diamond song 'Sweet Caroline' and then leads us all in a rousing, roof-raising chorus. We didn't win. I blame the Guinness …

MONDAY 24 FEBRUARY

On *Claire Byrne Live* tonight. We talked about COVID-19 and she got me to show people how to wash their hands. This has become a clear instruction from NPHET. Other respiratory viruses are spread from contaminated surfaces, so the risk with this one and others is that you'll touch a surface, then at some point put your hand up to your nose or mouth and infect yourself.

There haven't been any cases in Ireland yet, but even

so we're being told to keep our distance and wash our hands. I explained how soapy water and suds can kill the virus. I said how it dissolves the fatty bag that contains the genetic material for the virus – the RNA. I suspect it's the first time RNA has been mentioned on the show. Can it be that simple? If it comes here, and here's hoping it won't, will that keep it at bay? Keeping our distance is called 'social distancing'. A new term to learn. Social distancing and hand washing as the two main weapons against COVID-19: can it really be that easy?

TUESDAY 25 FEBRUARY

Two things always come in to mind on 25 February. The first is that this is the day my mother died, 38 years ago. The second is that it's George Harrison's birthday. Mind you, he's dead too. Very cheery …

Gave a talk in St Joseph of Cluny girls' school at the behest of Kathy in my lab, whose daughter goes there. I asked them at one point to name a virus and they all shouted 'coronavirus'! Went back into Trinity and signed some documents to release a lot of funding into Inflazome. The power I have as director! This will allow us to press ahead with our plans in the coming months.

WEDNESDAY 26 FEBRUARY

In London to speak at a Trinity UK alumni gathering.

My flight back to Dublin from Heathrow was delayed by two hours so I began working on my next column for the *Sunday Independent*. Found a quiet spot in a bar and got stuck in. I had prepared a piece about how rock bands are increasingly trying to go 'green' with carbon-neutral tours but then I thought, no – I'll write about SARS-CoV2 and COVID-19. I'll keep the green rock concerts one in reserve for the next time.

Writing about COVID-19 made me realise it could get bad. Shit. I did a lot of research. In China they have shut down cities, cancelled weddings, closed schools. I got the science in, saying how viruses were first seen in 1948 with an electron microscope. How SARS-CoV2 gets into our lungs via a key-and-lock mechanism – the key is the spike protein and the lock is called ACE2. I explained how, once the virus is inside, it acts like an unwanted guest who goes to your fridge, eats your food then goes to your spare room to have sex, making lots of little viruses. They then leave but blow up your house as they do so, irritating your lungs terribly. You then cough the virus out in droplets so it's important to wear a mask when you have symptoms, as recommended by the WHO.

I tried to reassure people though, reminding them that they have an immune system to protect them. I even mentioned antibodies and cytokines – I hope that's not too technical. And I've reminded them that there are already efforts to get a vaccine, and also treatments for those who get sick. I didn't tell them it can take years to make a vaccine. Or that we've failed to get treatments for other respiratory diseases. But we can hope. In the meantime, I've told them to call their GP if they're sick, isolate themselves, keep surfaces clean, wash their hands. I've told them to keep calm, remain vigilant and wait it out. I've said that this too will pass, just like SARS did. I hope so.

Also, a big announcement today. The rugby match against Italy scheduled for 7 March will be postponed. Things must be getting serious if they're postponing a rugby match.

FRIDAY 28 FEBRUARY

On the way into the lab this morning my phone goes on the Dart. It's *The Late Late Show*! They've asked me to go on tonight.

Stevie, my older son, told me he's been accepted into Cambridge to do a PhD in chemistry. My boy! So proud of him. Tremendous but a little bittersweet. How will he get on? It's a funny business, this parenting. You want the best for them and then you worry about them every step of the way. My thoughts turned to him as a little boy and how I used to bring him to school on my bike. He sat on a seat on the crossbar and when it was cold, he would put his hands over mine to keep them warm. And now … all grown up and heading to Cambridge.

The Late Late Show was interesting. I wasn't too nervous – too long in the tooth for that. But I felt a sense of responsibility to all the people who would be tuning in. In the green room I met big Niall Quinn. He's nearly as tall as I am! Another guest on the show handed me her phone and said, here's my husband, he wants to talk to you. It was John Jackson, who I had first met back in 1991 when I came back to Ireland from Cambridge. He was one of the few immunologists in the country at that time. Tonight he called to wish me luck. 'You're about to be the first immunologist on *The Late Late Show*!' he said. I sat opposite Ryan, with the audience in front of us. We talked about masks. The WHO advice was only to wear them if you had symptoms: they would trap the droplets and stop the spread. The key thing was travel: don't go to China. He thanked me for calming people down.

Back in the green room, someone said I should have said we should all start wearing masks, like they are doing in China. I said that wasn't the recommendation, but that I'd look into it.

On the way home in the taxi I read on Twitter that the WHO raised the COVID-19 alert to the highest level. Gulp. I wonder would that have changed my tone with Ryan? This is moving so fast. And it looks like China has effectively shut down. NASA released images showing air pollution has dropped dramatically over the country. The Chinese are certainly taking the virus seriously.

It's not every month we get a new virus name. Everything changing so quickly, and all the talk with immunologists is COVID-19. I fear it's going to get a lot worse before it gets better.

MARCH 2020

SUNDAY 1 MARCH

Off to the Bahamas for a conference on the immune system, metabolism and heart disease. Read a detailed piece in *The Economist* on COVID-19. Very sobering. It said the virus will test every government it comes up against. Another bit of scary news that the US and UK stock markets crashed this week because of COVID-19.

I land in Florida and all anyone can talk about is COVID-19. There is to be a huge music festival and they are wondering if they should cancel it. I fly on to Nassau (they pick these destinations to attract delegates). Met an old friend and scientific collaborator, Eicke Latz, from Bonn. He scared me. His colleague, a coronavirus expert, told him SARS-CoV2 will spread all over the world and won't stop until 70 per cent of the world's population is infected. And that in its wake it will kill a lot of old people unless we do something to stop it. I will remember the exact spot I was in when he told me this. Didn't know what to do with that information.

THURSDAY 5 MARCH

Flew from Nassau back to Europe and straight to another conference in Estoril, Portugal. Changing flights in Heathrow I watched a news bulletin in a bar, where the WHO Director General, Tedros Adhanom Ghebreyesus, said: 'This is not a drill … This is a time for pulling out all the stops.'

The conference is the annual meeting of the European Respiratory Society. All the talk is of SARS-CoV2 now. And especially at this conference, where all the European experts on lung diseases have gathered. During one of the scientific sessions, I saw on Twitter that Ireland has now had its first case of COVID-19, recorded on 29 February. A student in the east of the country who had arrived from Northern Ireland.

Then I got a phone call to tell me that someone in my lab was infected. What the hell, I said. Don't be ridiculous. One of my postdocs is from Italy. He'd gone home to Verona for the weekend, come back and developed symptoms. Had a test and came up positive.

My lab is now shut down and everyone has been sent home, including a transition year student, Fiona. This has to be seen as the height of irony. There's me in the media going on about COVID-19, and now someone in my lab has it. I am stunned.

As I sat down for dinner in a lovely outdoor restaurant by the sea in Estoril my phone went again. Someone senior from the HSE called me to brief me. I asked how long the people in my lab would have to stay in quarantine. She said, 'We don't call it that. We call it social distancing.' Then the Provost called to brief me. They were considering closing the whole building. Good God, is this the start? I hope my postdoc is going to be alright. I joined my fellow scientists and told them what was going on. Jaws dropped. They didn't know anyone who had been infected yet.

FRIDAY 6 MARCH

Back from Portugal. *The Irish Times* said there have now been 13 cases, four of them coming from Northern Italy, one of whom is connected to TCD. The first confirmed case was actually on 29 February.

Went for a drink in Fitzie's with Brian, my medic buddy. He was disturbed. Said he'd been hearing horror stories from a doctor friend in Bergamo, Italy. Crowded morgues.

SATURDAY 7 MARCH

Gig with The Metabollix, in the Dalkey Duck. We've been together now for 40 years … or so it seems. Began when we played in Dublin at a conference for immunologists in 2017. Bunch of scientists, medics and some real musicians, including our redoubtable lead guitarist, Chris. We've done loads of gigs at conferences but also in the Duck, where we've had a residency. The goal tonight is to to raise funds for Their Lives Matter, a charity in Dar es Salaam run by paediatric oncologist Trish Scanlan. It's for a cancer facility for children, a tremendous cause. We did a gig for them before in Kilruddery in Bray that had raised a fair bit of cash. But now we want to raise more to bring with us, because we're going to Dar es Salaam! Trish had asked The Metabollix if we'd be interested in going, and I'd bitten the bullet and said yes. Great cause, and the chance of an adventure to boot. So, I'd organised for us to go a few months ago, to do a fundraising gig there this coming St Patrick's Day and also for the Irish embassy there. It's going to be some trip!

Tonight was a mega gig. The place was heaving with people wanting to give money, and we raised over €2,500. We also rocked them to their very souls. Ciara Kelly sang with us as well – she is an occasional Metabollick. When the gig ended and the Duck closed, we headed over to Queen's, which is open late, and crammed into a snug there for more shouting and roaring. A splendid time was had by all. And no talk of COVID-19. Even I forgot about it for a while.

SUNDAY 8 MARCH

Tony Fauci, the lead immunologist in the US and Trump adviser, said that it might take 18 months to get a vaccine. Trump said he likes the sound of a couple of months better. But I've reassured people that a company called Moderna are ahead of the pack, creating a vaccine ready to test in humans in just 42 days. I also talked about chloroquine, a drug I worked on in my PhD, that is being tested, and a new antiviral called remdesivir. I'll keep telling people that science will save us. I don't know how long it will take, but we'll beat it in the end. We have to.

Phone call from one of the producers on *Claire Byrne Live*, asking me if I would ask the COVID-19-positive postdoc to go on the show on Monday. I called him up. He's reluctant but I said it would be a good thing. He could be anonymous and tell people his experiences. Reassure them. I've said I'll go on with him to help. He's even more reluctant at the thought of that.

MONDAY 9 MARCH

With Maura and Dáithí on the *Today* show. Great old

chat. About the virus, of course. I was going to tell them how a huge music festival had actually been cancelled in Miami and how we should cancel the St Patrick's Day parade. But just before I came on, the news came in that it had been cancelled. Now we all know how serious this is. A moment for the country to think about. That headline will go around the world: 'Irish cancel Paddy's Day'.

Also did the interview with Claire Byrne. If someone had told me a few months ago that I would be on *Claire Byrne Live* talking to a postdoc from my lab, whose voice was distorted, about him being infected with a deadly virus …

He explained the course of events – he'd called his GP and had a rapid test. Once that proved positive, he was immediately taken to hospital and is now in isolation. He told the ambulance man that he was an immunologist and the ambulance man asked if he knew Luke O'Neill. Gulp. It all sounded good – rapid response, treated very well. He said he felt fine and was bored. What a strange business that he's in my lab, the rest of whom are all still in quarantine.

Italy has gone into a national lockdown. Those scenes from Bergamo definitely scared everyone. They're saying the rest of Europe is likely to follow. What is going on?

TUESDAY 10 MARCH

Because of the upcoming gigs in Dar es Salaam, The Metabollix had a practice. We booked a studio in Dundrum with Eoin, a new drummer. He is sound. We ran through 15 numbers and it was good fun as usual.

Took a break in the middle and checked my phone. An important result for Inflazome came in. We had given a patient in Australia with a rare inflammatory disease called CAPS our new drug, and guess what? It worked. The drug stopped all the inflammation in his body. This rare disease is caused by a mutation in the gene for the protein NLRP3. The mutation makes this protein over-active and that is what causes all the inflammation. It can be a horrible disease – eye inflammation, sore joints, skin rashes. And all caused by the mutant NLRP3. He took our drug and within a day his symptoms began to disappear. Fantastic! And to think it started in my lab in Trinity. This bodes well for all the other diseases that involve over-active NLRP3, like Parkinson's and Alzheimer's.

I went back into the rehearsal room with a pep in my step and rocked out with the band, playing 'Johnny B. Goode' loud, Chris killing us with his lead-guitar breaks. As ever, playing gives me total joy, transporting me somewhere else.

Read an interesting report today on COVID-19. A choir that held a practice in the State of Washington led to over 80 per cent of those attending becoming infected. Good Lord. Yet another example of a high attack rate. This virus is spread by air, and it must be propelled out of people's throats by singing. Makes me nervous about that Dalkey Duck gig – such a huge crowd and lots of loud singing.

WEDNESDAY 11 MARCH

On Pat Kenny – update on COVID-19. Walking up to the Newstalk studio, two people recognised me. This hasn't happened before. Taxi drivers would often recognise my voice, as they have heard me on the radio with Pat. But this is different. Damn *Late Late Show*! I wonder will COVID-19 be the dominant theme for me and Pat? Surely not – won't people just be bored of it?

A grim milestone today: the first death in Ireland from COVID-19. An elderly patient in Naas General Hospital. Sent a shiver down my spine when I read about that. What will happen next? And the WHO has finally declared COVID-19 a pandemic. What took them so long? They said that it is now in 114 countries, with 4,000 dead and almost 120,000 infected. And there is no sign of it slowing down.

THURSDAY 12 MARCH

Interesting phone call with Dr Susan Evans, an

Australian doctor working on severe menstrual pain. She has evidence it might involve NLRP3, the target for our drug. She wants to collaborate with Inflazome and maybe even do a clinical trial. I said we'd be delighted. I love it when we collaborate with fellow doctors and scientists. She is convinced that blocking NLRP3 will help women, not just with menstrual pain but also with endometriosis, another painful inflammatory disease where there are few options.

I've started bumping elbows with people instead of shaking hands. Weird, but I guess this will be the way of things for a while. Got a taxi over to the Children's Hospital in Crumlin. Trish Scanlan had asked me to collect some boxes of chemotherapy to bring with us to Dar es Salaam. She often asks visitors to do that, as drugs are in short supply there. I got six boxes, medicine that will save children's lives. Better not lose them.

In the taxi on the way back into Trinity I heard on the radio that schools will be closing until 29 March. Good Lord. It's definitely getting serious. Rob, our keyboard player, called me to say he can't come to Dar es Salaam. He has young kids and needs to be in Ireland for them. Completely understand. I wonder should the band go after all?

Went on *The Tonight Show* with Ivan Yates. He asked if people can get COVID-19 more than once.

I said probably not. The immune system is supposed to protect you from reinfection. (The word 'immune' was actually coined from the Latin *immunitas*, which means 'exemption'. Soldiers who returned to Rome were granted *immunitas* from paying taxes. The word 'immune' was coined in the 1800s to describe how once people have certain infections, they are exempt from getting them again.) But it got me thinking: we just don't know for sure if having COVID-19 will protect you from getting it again.

Went home exhausted – and then saw that stock markets worldwide have suffered their greatest single-day fall since the crash of 1987. The world would appear to be in freefall. I thought of the Police song 'When the World is Running Down, You Make the Best of What's Still Around'. Can't help but feel that the world is indeed running down. Or, in the inimitable words of Marwood in *Withnail and I*, we are entering the arena of the unwell.

FRIDAY 13 MARCH

Woke up to the news that global stock markets have crashed. Appropriately enough, it's Friday the 13th. Not that I'm superstitious or anything, but this virus …

Today we leave for Dar es Salaam. Got up early. Got a text from the airline. The return flights had been cancelled. It said to go to the airport and either reschedule or get a refund. I thought about telling the band not to come, cancel the whole damn thing. But then I thought, *We have the chemo for the kids, the syringes, the cheque for €2,500 for Trish and the gigs for her to raise more money …* To hell with it. We're going.

Got to the airport and met the band. Discussed with them whether we should just not go. Went to the ticket office and they rescheduled our return flights, returning through Istanbul. The band are delighted – they've never been to Istanbul. And so we boarded the plane. As we took off, I felt a bit relieved for having got that far. The worries fell away as we took off. A long journey ahead.

SATURDAY 14 MARCH

Landed in Dar es Salaam at 3 a.m. this morning. Coming through customs, I was pulled over. The customs guy asked me what was in all the boxes I had in my big suitcase. I said drugs and syringes, then paused for effect to see what he would say. I added that they were for the children's hospital in Dar es Salaam and showed him a letter that Trish had given me. He laughed loudly and let us through.

The Irish embassy had sent a driver. We felt like rock stars. He drove us to the hotel. I fell into bed but the rest of the band went for a swim. Got up at 11 a.m. and went down for breakfast. Trish was there. She was happy we'd come but looked worried. The Tanzanian government had issued an order to limit gatherings to 100 people. She's had to curtail the number of people coming to the St Patrick's Day ball and moved the venue to the rooftop of a friend's house. She said it would be fine. Her friends were planning all kinds of Paddy's Day decorations.

We spent the day hanging around the hotel, swimming in the pool, drinking cocktails. Had a bit of a laugh with Trish remembering how in early January we had met up to discuss the plan when she was back in Dublin. Me, Brian and Chris gate-crashed a dinner in a vegan restaurant where she was meeting up with her old clinical pals from Crumlin. They are a great support for her in Tanzania, helping with diagnosis and sending supplies. We weren't too impressed by the vegan food, although I'm sure it was top notch. I had chili sin carne and Brian had BBQ cauliflower, which tasted like cauliflower with BBQ sauce. We went to Burger King after for Whoppers …

The weather was hot and humid. The pale Irish skin of the band looked very fragile in the Tanzanian sun. Lots of sunblock. Had to leave the poolside to do a podcast interview with David McWilliams, who couldn't quite believe I was dialling in from Tanzania. At 7 p.m. we headed to the venue for the gig. The PA was delivered and set up. We did a soundcheck. Eoin says the drums weren't great and Chris was horrified by the electric guitar he was given. Without Rob, we had no keyboards to fill out the sound. But, as Chris said, as long as we have my right hand pushing out the chords, we'll be fine.

It was boiling – I'd say the temperature was about 90°. It was like playing in Vietnam ... not that I've ever played in Vietnam. We got them all up dancing. In the middle, we presented Trish with the cheque and the meds. We played a bit of everything, getting them on the dancefloor with 'All Right Now' and finishing up with 'It's a Long Way to the Top'. The sweat kept getting into my eyes, stinging them, and Chris broke three strings. But we did it.

This was our setlist:

ALL RIGHT NOW (A)
USE ME (EM)
OH DARLING (G)
BE MY BABY (E)
OLD TIME ROCK 'N' ROLL (G)
TWIST AND SHOUT (D)
TREAT HER RIGHT (E)

Fifty Ways to Leave Your Lover (Em)
Pencil Full of Lead (D)
Preacher Man (E)
Valerie (Eb)
No Diggity (F#)
Honky Tonk Woman (G)
Stormy Monday (G)
Superstition (E)
Play That Funky Music (E)
Let's Dance (E)
Back in the USSR (E)
Brewing Up a Storm (Am)
It's a Long Way to the Top (A).

We went back to the hotel, with that after-gig glow. It's such a buzz! Playing loud music to a crowd is just something else. Went to our rooms to freshen up. I got back down to the bar first and ordered a gin and tonic.

And then a text came in from Brian. Two people who had been down in the Dalkey Duck the previous Saturday had tested positive for COVID-19. What the actual fuck? The band assembled and I told them the news. Nervous laughter. I knew immediately what we had to do. 'Right,' I said, 'we have to get back to Dublin as quickly as we can.' Talk about buzzkill. The guys were looking forward to a night on the town. I said any of us might be infected and we might infect people here.

I went on to SkyScanner on my phone and within five minutes had booked five tickets from Dar es Salaam to Dublin via Doha the next day. I had visions of huge hassle trying to get flights, so the relief was enormous. I told Trish, who said she would send over some N95 masks. I'd never heard of N95 masks. I knew we couldn't meet her again. If she were to get infected and bring it to the hospital that could mean tragedy for the sick children.

The band called their families with the news. Eoin's girlfriend runs a pub in Temple Bar. When he came off the phone he told us the pub was closed, as are all the pubs in Ireland. My jaw dropped. Pubs in Ireland closed? How can that be?

We lounged in my room in quarantine. The lads played a bit of music. Taylor, our bass player, was told not to fart.

MONDAY 16 MARCH

We got a taxi to the airport in almost complete silence. Masks on, we checked in. The airport was empty. I saw a headline on a big screen: 'Dow Jones Industrial Average has the single largest point drop in its history.' Feels like the world is falling apart.

The flight from Dar to Doha was almost empty. We had to leg it to the gate for Dublin. There was a guy waiting for us to get us through everything as quickly as possible. It felt like we were leaving Saigon after the Vietnam war. We passed loads of duty-free shops full of luxury goods. It was so quiet. I thought, *Is there an apocalypse coming?* On the flight to Dublin we had four seats each so we could stretch out and sleep. No food was served and everyone wore masks the whole time.

WEDNESDAY 18 MARCH

Landed back in Dublin. My Twitter has gone crazy. I sent some tweets about COVID-19 and hundreds and hundreds began following me. Over 10,000 now, up from 2,000 when I left for Dar es Salaam. Coming through the airport we were quizzed about where we had been. When I said Doha, the woman said we need to self-isolate for two weeks. I said, 'What?' She took my phone number and gave me a leaflet.

I got a taxi home. The taxi driver told me how the country was now in lockdown. *Isn't quarantine good enough?* I thought. Are we in some Hollywood movie? It certainly sounds like that. Shops and schools are closed, and people have been told to stay home. I told him I would have to self-isolate for two weeks because I'd been in Doha. He edged away from me towards the windscreen.

He said how the Taoiseach, Leo Varadkar, had made a speech on 17 March during his US trip for St Patrick's Day. People are saying they will remember where they were when they heard it. I checked it out on my iPhone. He said how today's children will tell their own children about how there were no St Patrick's Day parades in 2020. He said how Ireland was in the midst of a global and national emergency – a pandemic. How cases would rise to 15,000 by the end of the month and then even more beyond that. Large public gatherings cancelled; pubs shut; curtailing of weddings. How elderly people would have to stay home, but that they would have food supplies – he said this would be called 'cocooning'. He reminded everyone how not all superheroes wear capes – some wear scrubs and gowns. How it would be a shared enterprise of all humanity that finds a treatment and a vaccine that protects us. The words made me feel a chill. I'll remember where I was when I read them – in a taxi driving from Dublin Airport. It will be some time before I head back to the airport, who knows how long.

I checked my phone for what was happening in Tanzania, fearful that I was patient zero. Some cases had been reported in the north of the country. None in Dar es Salaam yet. I hope things will be alright there. Trish had said the health system is so bad that there could be a huge death toll if COVID-19 struck. She said the one thing the country had going for it was the young population. It's clear now that COVID-19 is mainly a disease of people over the age of 65. I'll watch it closely.

Got back to the house at 4 p.m. Life at home begins.

THURSDAY 19 MARCH

Still exhausted after the Dar es Salaam trip. But – Stevie came home today from Edinburgh, because of the lockdown being imposed in Scotland. So great to see him! Joy unbounded. *Prime Time* called, wanting me on. I said I have to stay home. They said they could film me in the garden. That's how weird things are now. I'm happy to give information on what's going on. I think we scientists have to step up now and play our part. They filmed me a good distance away. Asked me about where we're at with COVID-19. I gave the update: Ireland now has over 500 cases. The instructions from the government are for everyone to socially distance, stay home, wash hands. I said we can beat this together and that the science was ramping up to get a vaccine and treatments. I said the amount of science going on is unprecedented. I hope that people will feel less scared. I really believe science will beat this, but I know there will be bumps along the way and I don't know when.

This makes me a little nervous, like everyone else, but then I remind myself of how great science is, and then I don't feel so bad ... probably because science is one of my favourite things.

Meanwhile, the Chinese are reporting the start of vaccine trials. That was quick! It's all very uncertain. We never got a vaccine for the common cold, which can be caused by a similar virus. But progress was made against SARS, although a vaccine was never approved. What makes us think that COVID-19 will be any different?

Saw Tedros Adhanom Ghebreyesus on the news and he laid it on the line: 'Once again, our message is test, test, test.' Simple, right? Surely we can use testing and tracing to get this virus under control?

The Grand Ole Opry in Nashville played to an empty theatre – the first time since 1925. Must have been very weird.

SATURDAY 21 MARCH

This media thing has taken off like a rocket and I am just about holding on. A lot of people asking me – seven requests today. RTÉ Radio One. 98FM. A bit of pressure, as I have to make sure I get the facts right. But it's a privilege to be asked and I want to help people. Feel a sense of utter weirdness.

Following stuff on Twitter and in the media about

COVID-19. It's spreading like wildfire. Such stress out there. And then there's the conspiracy theorists and people accusing the experts of scare-mongering, exaggerating everything, claiming it's a 'casedemic'. Global cases have reached 250,000 with 10,000 deaths. I believe these facts.

A woman sent a nasty email, saying she'd seen me on *Prime Time* and that I was just too offhand about such a serious thing. A man emailed to say I had only one agenda, and that is to scare people out of their wits. Another person tweeted to say I was only doing it for the fame. It can be hard to be forgiving of nastiness, but I think it illustrates that people are scared.

Tony Fauci was on CNN. I met him once, in 2009. I was the 'Immunology Guru' at the National Institutes of Health, where Tony heads up the Immunology Institute. We had a great chat. He said something about how scientists should engage with the media. Try and base what you say on data. If you don't know something, just say it. And remember – it's not about you, it's about what you're trying to explain in as clear a manner as possible. I must remember these three things!

Netflix and YouTube have reduced their video quality in the EU to prevent internet gridlock because so many people are at home. McDonald's is going to shut all its restaurants in the UK. Oh, and the Olympics have been postponed to summer 2021. Things should be well over by then. Hold on to that thought. And there are lots of reports of people raiding supermarkets. Jokes about supplies of toilet roll running out. I guess there's something of a herd mentality, with people being afraid of scarcity.

TUESDAY 24 MARCH

Today, Ireland went into full lockdown. Everyone staying home. Do not go beyond two kilometres. No social mixing outside your own bubble. Yet again, what the fuck?

WEDNESDAY 25 MARCH

Did a Science Gallery podcast today. I've been asked by scientific journals to give updates on what we work on in my lab. They're called reviews. I've divvied them out to my lab as it will give them something to do as they are all still at home. I worry about them. They need to progress their PhDs and build careers. How will they manage? The reviews will cover what we're working on in the lab every day. A big topic is how cells of the immune system use nutrients in different ways to other cells, especially when they are in overdrive. And how we might stop that to help treat inflammatory diseases. And of course, how my favourite inflammatory protein, NLRP3, is such a great target for anti-inflammatories. Scientific journals can't get enough of that.

Gave a talk by 'Zoom' for the Royal College of Physicians on COVID-19. It's just not the same. Can't get the buzz off the audience. Don't like it, but I did it. Mary Horgan is president, and she gave a good talk too.

Was supposed to be at the Dutch Society for Immunology today, giving the keynote lecture of the conference. It was to be held in a town called Noordwijkerhout. Cancelled. All the conferences are being cancelled one by one. My actual world closes in, but my online world goes into hyperspace.

A lot of talk about hydroxychloroquine as a therapy for COVID-19. Trials are running. This is reasonable, as it has been shown in the past to be an anti-inflammatory and have antiviral properties. We'll see. It's not as if hydroxychloroquine was ever approved for another respiratory virus. But we need as many shots on goal as we've got. It's a phrase I keep using in the media too. Hope people don't get sick of it.

And the death toll in Ireland has now hit 19.

SUNDAY 29 MARCH

The Taoiseach has announced a national stay-at-home order. In the top right-hand side of the TV screen it says 'Stay at Home' or '*Fan sa Bhaile*'. Unprecedented.

Even during the 1918 pandemic there was no order for people to stay home, which was one reason for the high death rate then.

Lot of emails thanking me for giving updates. I'd say it's 90:10 positive, which is good. I would stop doing it if I felt it was counterproductive or not useful. Such an appetite out there for information.

Stevie has been accepted into Gonville and Caius College. It's the one that was in *Chariots of Fire*! Wonder if he can run fast?

We had a big Zoom call in Trinity today with all my immunology colleagues. We discussed seeking funding from Science Foundation Ireland for a centre to work on COVID-19. My colleagues Kingston Mills and Aideen Long will co-lead it. All our labs have something to offer. My lab will work on itaconate, the natural brake we found on the immune system. We'll now examine its antiviral and anti-inflammatory effects. There were a couple of discoveries last year showing it could kill Zika virus and also influenza, so you never know. No lab work at the moment, though, as the labs are still shut. Bah.

Read about the neurological features of COVID-19 – loss of taste and smell are common. This happens with other respiratory viruses, but it seems especially prevalent. Sent a shiver down my spine. What if this one damages the brain? So much we don't know still.

TUESDAY 31 MARCH

Some good news today. Sitryx, the other company I co-founded back in 2016, has signed a deal with Eli Lilly for $50m. This will give Sitryx lots of funds to explore new anti-inflammatory drugs that we are trying to develop. Sitryx might also be involved in our COVID-19 work, which would be great. They've expressed a strong interest in it. Having them on board would be superb, as they bring a lot of expertise and might help take any discoveries we make to patients, which of course takes time but is the ultimate goal of all this.

Approximately 2.6 billion people (one third of the world's population) are in some form of lockdown. Good Lord. Working from home is a challenge. Spending all day everyday looking at a screen. If this virus had come along even 10 years ago this would not be possible. I guess people would have just gone to work, and we'd have had to accept a level of death and illness as immunity built up and we waited for the vaccine to arrive. Read about how in previous pandemics the workers still had to work in the fields to keep the food supply going, spreading infection as they went. A new virus keeps one third of the world's population at home on their computers. No one saw that coming now, did they?

Another staggeringly weird month is over. The heat of Dar es Salaam. The media frenzy. The virus, the virus, the virus.

APRIL 2020

WEDNESDAY 1 APRIL

In the words of Paul Simon, April, come she will. At least the weather is looking up. The two weeks since Dar es Salaam have elapsed. I can go to the shops to buy food!

On with Pat, by Skype as usual these days. We discussed antibodies to COVID-19. A lot of interesting questions, like how long can antibodies last for? More work has been done on this question and they do seem to wane, but the cells that make them – the B cells – can persist and are ready to be triggered once reinfection happens. I reminded everyone how antibodies are the workhorses of the immune system. They are tailor-made for fighting whatever the infectious agent is and can neutralise it. For some diseases, the cells that make them can last a lifetime, protecting us from disease. For other diseases they are more short-lived, so we just don't know with COVID-19. If you have COVID-19 once, can you get reinfected? We still don't fully know yet, but it's likely that there will be some protection. After all, that's how the immune system is supposed to work. What strikes me is that science is happening in real time, and I'm telling the public about it. Usually I only talk about science that is definitive, so it's a bit of a departure from that. I hope people are appreciating more how science works.

Tony Holohan, our Chief Medical Officer, who has been keeping the country going, was admitted to hospital tonight. Non-COVID-related, but unnerving for everyone. I hope he's going to be OK. It sent a shock wave: Dad's in hospital.

THURSDAY 2 APRIL

Global cases hit 1 million today. And all from a single person infected by a bat? How can that be? A very contagious virus that spreads from people who don't even know they're infected, that's how. There are of course lots of other examples of how viruses jump from another species into us, causing disease. The virus that causes AIDS – HIV – came from a chimpanzee. Influenza usually comes from birds or pigs. And, of course, with the related viruses SARS and MERS – it's thought that SARS came from civet cats, while MERS came from camels. A strange business indeed.

Striking image on the news today: the US navy hospital ship *Comfort* docked at Pier 90 in Manhattan. It has 1,000 beds. They have started to fill up.

And then I was interviewed by MTV news. This was set up a couple of days ago. MTV at last! Couldn't get enough of it back in the '80s. I should have brought The Metabollix.

FRIDAY 3 APRIL

An interesting paper has come out suggesting that in countries where BCG vaccination is used (Ireland isn't one of them), it might give some protection against COVID-19. This is right up my street, as I've done some work on BCG with Romanian-Dutch professor Mihai Netea in Nijmegen. It's the TB vaccine but it can give a general boost to the immune system (called trained immunity), which might protect against other infections. It's been shown to protect against influenza for example, so you never know. The study is just an association though, but clinical trials are planned for BCG and COVID-19, which would be great, as it's a very safe vaccine. Spoke about all of this on Newstalk.

Two brief interviews with *The Irish Times*. Also

recorded a talk on immunology for RTÉ Junior, who are running lessons for kids stuck at home. Home-schooling is driving some parents nuts! Another pressure on people. I was so glad to do this, though – teach them immunology early and who knows, some may become immunologists. And it might give the parents a break too!

The Centers for Disease Control and Prevention is recommending that people wear face coverings in public. About time too. Ireland should do this – immediately. The evidence that masks help has become more robust, mainly because the virus comes out in small droplets, and masks can trap them. Seems obvious, but many disagree, and it's become highly politicised in the US. How can a mask become political? Maybe seatbelts were at one time too.

Ireland now has over 4,000 cases, with 120 deaths.

SUNDAY 5 APRIL

Another Sunday. Fed up. Had a big piece in the *Sunday Independent* on the mysteries of COVID-19. Children don't get sick from it, which is great, but it's not known why. Also, not everyone in a household gets sick from an infected person. Could it be due to the dose of the virus, with not everyone getting exposed to the same dose? And why are men more susceptible to COVID-19 than women? Wrote about a report saying it is three times more dangerous than flu. It's important to remind people of that, as some are still saying it's 'just another flu'. Again, go by the data. Please go by the data.

MONDAY 6 APRIL

Mihai was on with Pat this morning, describing his BCG work. Good old Mihai – both an old friend and a talented infectious diseases doctor. He's from Transylvania, which I always find amusing. If I had COVID-19, I'd want him to look after me, but I would not let him near my neck.

A reporter on the BBC just announced that Boris Johnson is in intensive care with COVID-19. This virus can infect anyone and make them very ill. He will have great care, so he should be fine.

TUESDAY 7 APRIL

A very busy day, as if there are no others. Started with a call with Inflazome. We're answering more and more questions from Roche. I've never seen diligence like it. Seems like an age since we met them in San Francisco in January. And yet it's only three months. Funny how we perceive time.

Then I took part in an online conference with an organisation called 'Access to Medicines'. They are all about trying to ensure that medicines get to the people who need them most, and this will include treatments and vaccines for COVID-19. I also had a call with the Trinity COVID-19 team. We discussed how to structure the application being made to SFI. Then a call with Marjolein in Leiden. She is part of the viruses/immunometabolism EU collaboration. We discussed experiments we might do with itaconate and SARS-CoV2. We're going to start doing experiments on COVID-19! So exciting to be part of this big international scientific adventure. Hauke is the German student in my lab who is part of the EU collaboration, so he will start doing experiments. Alex is also keen to help. They will be great to have on it. Like so many immunology labs all over the world, we turn to working on COVID-19.

I was on *Ireland's Call*, a special RTÉ programme on COVID-19 running twice a week. This was exciting, because they sent a taxi. I haven't been out in weeks! It had a big Perspex screen separating me from the driver. I took a pic and tweeted it: 'Driving Miss Daisy.' The programme is being made from the Department of Health. Strange, quiet atmosphere there. And I met Tony Holohan! Good to see him. We had a brief chat. Such a sound person. Yet again, in Ireland we're blessed to have people like Tony.

The Taoiseach has announced that the measures that came in on 27 March will be extended until at least 5 May. The lockdown continues for at least another six weeks. But Wuhan has reopened, so they've managed to get the virus under control?

THURSDAY 9 APRIL

The Twitter frenzy continues – 20,000 followers now. Had my first Zoom lab meeting. Kind of worked, I guess. I asked them all to tell us something enjoyable they had been doing, just to cheer them up.

SATURDAY 11 APRIL

I started correcting the proofs of *Never Mind the B#ll*cks*. I'm dropping in mentions of COVID-19. It's pretty easy, as the book is all about how science is such a great lens through which to view the world. I laid into Trump in the introduction for being anti-science. He recently said in a press briefing that it might be an idea to inject bleach into yourself to kill the virus. I mean, FFS. Also added info into the vaccines chapter. Still not fully sure how the book will go down, but hey, people might like it, given all the attention on science right now.

Had a bit of respite today too. Brian came over and we had coffee in the garden. By God, we'd love to go back to Fitzie's for pints like in the good old days! Will they ever come back? Of course they will. It's only a matter of when.

SUNDAY 12 APRIL

Went to see the mother-in-law today in the nursing home. Desiree was her usual cheerful self. She is a great woman and I'm so fond of her. Always has a smile. This was a 'window' visit, as no one is allowed into the nursing home because of the risk to older people. We chatted on the phone, through the window. Her short-term memory is shot, but she still recognises me. She knows about the virus, so we had a chat about that for a while. I was longing to go in and give her a hug. She looked so fragile, but I'm always amazed at how cheerful she is. She was more worried about me than I was about her. Walking back from the nursing home, two people recognised me in the People's Park in Dún Laoghaire. This is now the way of things.

Podcast with David McWilliams. We discussed all things COVID as per. He drew a parallel between the immune system and the economy. He might be on to something!

MONDAY 13 APRIL

Was on *Sky News* this morning. The Irish reporter came over to the house at 8 a.m. and we went down to the seafront to film. As luck would have it, I cut myself shaving. Sod's law. So my sister Helen, who watched it, said, 'What was that big clot on your face?' Thanks, Hellie …

Tragedy struck! The dishwasher is broken. I predict this will tip us over the edge, what with all of us locked down together. We need an emergency plumber to make sure we don't kill each other. I'll tell him it's a matter of life and death. We're lucky, though. We can all be in separate rooms on Zoom. This avoids me being embarrassed by Sam walking into shot in his underwear.

Read a good account of the disease profile in Ireland. Eighty per cent are mild to moderate, 14 per cent have severe disease and 6 per cent are critical. Much more serious than flu, which is a constant comparison in the media. Analysis also indicates it's worse than flu across all age groups. Hard to get an exact fix on it, because it affects older people more severely. But from my reading of it, overall mortality across the population is around 1–2 per cent, whereas flu is 0.1 per cent. I have to wonder, these people who deny the seriousness of COVID-19, is it classic denial? We don't like to be told negative things about ourselves, so we fight back. The level of aggression usually correlates to the level of denial, in my experience anyway. Maybe that's what's going on here.

TUESDAY 14 APRIL

Was on *Off the Ball* on Newstalk, and a clip of what I said of how important sport was for people's mental health was used on the news. That's the first time that has happened: 'Immunologist Luke O'Neill said ...' I emphasised how exercise is so important to keep us going, both mentally and physically. With gyms closing, people should find other ways, be it a run or even a brisk walk. I said how it was great for your immune system too, as the immune cells in your blood slosh around and wake up when you exercise.

Taxi into Virgin Media One for *The Tonight Show*. Ivan in top form. A guy called Michael Levitt was saying that the virus will soon burn out and then people will be immune. He is an eminent scientist, having won the Nobel Prize in 2013 for chemistry. I was critical of what he said, since there's no evidence – I said he was talking rubbish! This is not the thing to do if you're an academic. Always be moderate, because in that way you can reach the truth. So I felt bad about saying it. But even though science is all about dispassionate analysis of data, scientists can get angry and upset too. I just can't see this virus going away any time soon. We know so little about it, so we must err on the side of caution.

What Levitt was referring to was the population reaching herd immunity: when a sufficient number of people are infected and become immune, the virus has not many places to go and dies out. We have no idea of how long this will take or even if it's possible for a highly transmissible virus. The job of vaccines is, of course, to build up immunity safely in a population, but to try and do this naturally risks carnage, as it won't be possible to protect all the vulnerable from severe disease and death. It's an argument that will come up again and again, I predict, mainly from people who don't know how the immune system works.

Donald Trump announced that the US are pulling out of the WHO. For crying out loud. In the middle of the biggest pandemic in 100 years. What is he thinking? Not much, I would guess.

WEDNESDAY 15 APRIL
A 23-year-old was reported to have died from COVID-19. Not just a disease of older people. And two healthcare workers have died. Why are many still questioning the seriousness of it? Let them work on the front line and then they might change their tune.

THURSDAY 16 APRIL
I got a grant from SFI for my research into itaconate,

which is great news indeed. That will fund four people for the next five years in my lab. Superb!

Did a 30-minute debate on Euronews. A long time on air. Going global now. At least not many in Ireland watch Euronews.

Good analysis from NPHET today. The lockdown is working, driving down the growth rate. They've said it's close to zero. Let's not blow it now.

Watched a movie tonight: *The Damned United*. It's about Brian Clough's time as manager of Leeds. I remember that! I really enjoyed it – got completely lost in it. So important to be able to do that from time to time.

SUNDAY 19 APRIL

Was on the Brendan O'Connor show on RTÉ. It was good. Met Stefanie Preissner. She asked if I'd do a podcast with her, and I said I'd be delighted to. I must admit I don't really know Stefanie. She seems very sound. Went to see Desiree again with Marg and we then off to Lidl. Got recognised again. Read Brendan O'Connor's piece in the *Sunday Independent*. He's spot on and very witty. He wrote about how when he's feeling down he runs/talks/reads his way out of it. And also how it's important to be positive for the sake of his family. That's always been my way too. No point in laying it on them. I was always conscious of that with my two lads. Don't lay your bad mood on someone else and make their day tougher. It's been a kind of principle of mine. I think it stems from when my dad was depressed and all he did was complain to me the whole time and it made me feel bad. I remember once visiting him in the nursing home he was in. One long moan-fest. I was walking back down the driveway lined with trees after the visit promising that I would never do that if I had kids. Not blaming him, as he was in such pain. But I figured – how about I try and not do that, being aware of the effect it was having on me? With COVID-19, even though things are bad, I try not to sugar-coat it, and always look for a way out.

MONDAY 20 APRIL

Ed Kenny, my cousin's son, asked me to send photos of my dad's war stuff. Funny to have my dad so prominent in my mind these days. These are my most treasured possessions. It was stuff he kept from World War Two. The manual for the Sherman tank he drove. His campaign medals. It was good digging them out and Ed was very appreciative. I think so many of us are reflective in these troubling days. We think back to the past and try and get some comfort from that. I got that today sorting through my dad's old war stuff as he himself had shown them to me, so proud of what he'd done. Ah,

Dad – I sometimes long for those days when we had good times together.

On *Claire Byrne Live.* They had mocked up a supermarket and we used wipes on a shopping trolley and explained how there's no need to wash down shopping. Marg and Stevie said I shouldn't do those kind of demos – they're embarrassing. As I quoted in *Never Mind*: 'Behind every great man is a woman rolling her eyes.' But I don't mind at all. It could be helping people.

WEDNESDAY 22 APRIL

Stuck at home. On the radio all the talk is of how people are coping with lockdown. Everyone is baking sourdough. And it's so quiet, especially at night. I go out into the garden and there is a such a beautiful stillness. You can almost feel the Earth moving. And then in the mornings, the birds singing in the trees. There's some kind of finch in our front garden and it flits into the bushes, squeaking like a child's bath toy. I love it.

I am finding this media stuff surreal. Today was media-heavy. I tell Pat and the listeners about a drug called ivermectin, which was discovered by an Irish scientist called Bill Campbell. He had been searching for anti-parasitic drugs to use to treat farm animals suffering from parasites such as roundworms and liver fluke. And he found one! It was sent to him by a Japanese scientist called Satoshi Ómura, who had collected soil from a golf course near Tokyo, which had a bacteria in it that could make an anti-parasitic chemical. Bill purified it and modified it to make ivermectin. This drug was highly profitable for Merck, but then Bill had an idea. Why not use it to treat parasitic diseases in humans? He had one in mind. A disease called river blindness, caused by a parasitic worm infecting the eyes, mainly of children. Bill showed ivermectin works against that too. He convinced Merck to give it away for free to poor countries where river blindness is endemic. And in doing so, Bill stopped millions from going blind.

For this, Campbell and Ómura won the Nobel Prize for Medicine in 2015. He is the only Irish winner of this Nobel Prize, and it's so well deserved. We're proud that Bill studied zoology in Trinity in the 1950s, where, he said, Professor Des Smith inspired him to work on parasites. It's a great example of how a teacher can inspire someone in their life, and Bill always gave Des great credit.

A report has come in from Australia suggesting that ivermectin might kill SARS-CoV2. Trials will be starting. An example of serendipity.

THURSDAY 23 APRIL

A momentous day. Because we have projects working on COVID-19, I asked permission from the Dean of

Research in Trinity, Linda Doyle, to grant us access to our lab. This is permitted if you can justify it, so I listed Zbigniew (who we call Z), Alex and Hauke. All of us are mad keen to return. We could be back in action! One of the great things about working in a lab is a sense of camaraderie. We have a shared goal – to make scientific discoveries that might help people. A shared mission is a great thing for us humans. We evolved to revel in it. And we share all the successes and failures too. This damn pandemic has stopped all that. Can't wait to see them again.

Wrote what I think is an important piece for the *Sunday Independent.* We should all be wearing masks – commonplace in Asian countries like Japan and China. And the evidence that they are useful has grown in the past few weeks. I read an analysis by Oxford scientist Trisha Greenhalgh and some of the scientific papers on this, and I'm convinced. Hardly anyone in Ireland is wearing them. This is part of what I sent in – worth recording!

WEAR A COTTON FACEMASK

Remember the phrase 'Say it don't spray it'? Well, sadly it doesn't apply to COVID-19. Three important studies have just been published that say everyone should wear a simple cotton facemask when they go out, especially when they go to supermarkets or on public transport. Cotton masks have been shown to decrease transmission of viruses 36-fold. That's a lot. The evidence is now so strong, the Irish government should secure a supply (as the South Korean government did) and a law should be passed to ensure compliance.

What is this new evidence? Well, yet again there has been a frenzy of scientific activity. Studies from as early as 1934 have been dusted off. Special laser technology has been deployed. And last week, three important studies were published. Scientists showed that when we speak, we produce tiny droplets, called aerosols. These can float on the breeze and travel far, well beyond 2 metres. Scientists showed that these aerosols are likely to carry the SARS-CoV2 virus. And a major study showed that as many as 44 per cent of people become infected by someone who doesn't have any symptoms. These three scientific facts say one simple thing: everyone should wear a cotton facemask when they go out, especially when they are in places where there are other people, such as shops and supermarkets or on public transport. This is not to protect you from infection. It's to *protect others from being infected by you. Like drink-driving laws, it's to stop you harming others.*

The official advice on facemasks, still being given by the HSE and the WHO, has been that you should only wear a cotton facemask if you have symptoms. This made sense before this new information came to light. Now that we know that you can have COVID-19 without symptoms, and that just by speaking you might be spreading it, and that the aerosol can travel beyond 2 metres, you must wear a facemask. It is now essential that the guidelines (and even the law) change. If it doesn't, it would be as stupid as ignoring the evidence that seatbelts save lives.

The US, China, Japan, Germany, France, India, Brazil and Canada all say wear facemasks. Trump, as ever, blithely goes his own way, and said he wouldn't wear one, apparently because he meets dictators a lot and for some reason felt they need to see his face. Hey Donald, how about you get like a cowboy and pull a kerchief over your face? You can bring your toy gun and holster to the meeting too. Wearing facemasks is actually a badge of honour in many Asian countries. It says, 'I'm looking after you, and you're looking after me.'

Austria and the Czech Republic both imposed a lockdown, but the Czech Republic also made the wearing of masks mandatory. The curve flattened there much more quickly than Austria. Some weeks later, Austria recommended facemasks for all, and guess what? The infection rate fell. In every country where facemask use was made law, or where facemasks were provided to all citizens, the infection rate and deaths fell.

You might be wondering why the hell hasn't our own government (through the HSE) changed its guidelines on this. There have been studies arguing against everyone wearing facemasks. There is hardly any scientific evidence to support these concerns. So go on, make your own facemask. Even using a bandana or kerchief will help. You will be caring for others and yes, the science says it all: facemasks will help in hastening the end of this lockdown.

I wonder will it make any difference? I sent it off around 11 p.m. and made a cup of tea. And then something strange happened. I had the radio on, and a song came on: 'Matchstalk Men and Matchstalk Cats and Dogs'. Corny as hell. It's about the painter L.S. Lowry, who painted scenes from Salford in England.

My dad grew up in Salford, and he used to sing the song, because it was about his hometown. And you know what? I started crying and I spoke to my dad, which I do from time to time. He didn't answer back, of course, because he's been dead for 24 years, but hey, maybe he answered in my brain. Music can do that to me sometimes – unleashing an emotion. Jeez ... the emotional lability of these times. It felt good talking to Dad.

FRIDAY 24 APRIL
A red-letter day! A day of days! I went back into the lab. The Dart was very, very quiet. I'd say six others in the carriage. I was wearing a cotton facemask that I had made for myself out of an old T-shirt, by following a YouTube video. It was stapled together. Crude but effective. I felt really strange wearing it. Like the Lone Ranger.

Momentous day as there's the first report of a vaccine protecting against COVID-19 in animals. We might get out of this yet. And Oxford injected humans with their vaccine yesterday. They had been developing a vaccine from back in February, and had seen evidence that it might work too in a laboratory setting. We are under way. And the Chinese data, although limited, says it might work. However, a vaccine working in animals doesn't mean it will work in humans, but it's an important first step. Oxford will first of all make sure it's safe in humans, as that is as important as anything else. My fingers will remain crossed for quite some time. The UK death toll has now passed 20,000.

In the lab we act like children going back to school after the summer holidays. We had a big meeting, socially distanced. The other three asked why I was wearing a mask.

We planned some experiments. I drew on a white flip-chart and then stuck it up in the lab. Our work will be all about itaconate. We've been working on that since 2017. It's like the off switch that gets flipped when the danger is gone. Danger can be an injury or caused by an infection. We had shown that itaconate can protect mice from dying of septic shock caused by bacteria. Then another lab showed it could protect mice from death from Zika virus. That got me thinking, *might it work against COVID-19?*

The team are on it. There is much excitement. They are delighted to be back in the lab and actually doing something. And most importantly, working on the greatest threat to our health in 100 years. I could see the zeal in their eyes. They were burning. All our eyes were burning with zeal. This is what all those years of training and research are about. A chance to make a big difference with the knowledge we've built up. We were all high as kites with the excitement of being back.

Headed home at six. Friday evening and Pearse Dart Station was completely empty. It's like the zombie apocalypse. I kept an eye out for lurching zombies. It felt good to be out, though. No one at all in my carriage. Z, who is from Poland, said it was like we were the first workers to go into Chernobyl to clear up the radioactivity. Not a bad analogy …

At 7.30 p.m. I gave a Zoom talk to the sixth-year biology class in St Columba's. Got some great questions off the students. Humphrey Jones asked me. He's a great biology teacher, and was a huge help with my children's *Irish Science* book. Sales have started climbing again. I guess people are ordering it online. One mother even emailed me to say she had wrecked her kitchen with some of the experiments.

Had a great jam with Sam afterwards. He was on keyboards, and I banged away at the guitar. We tried some Elbow numbers and it wasn't too bad!

Went to bed tired but satisfied. Some day. And some days of adventure ahead, I feel.

SUNDAY 26 APRIL

Tony Holohan reported this week that 45 per cent of deaths so far were in nursing homes. Could we have done better to protect those vulnerable people? There may well be a public enquiry into that. Probably the only way to protect them would have been if all the staff stayed in the nursing home and never left. Difficult for them. My thoughts turned to Desiree. What if one of the deaths had been her? Her home is doing a great job, thankfully. But it's all so sad.

These thoughts don't help my mood on another Sunday, Bloody Sunday. It's like *Groundhog Day*. Every day, every week blends into the next, in spite of all the different things I'm doing. A big difference is no travel. I'm supposed to be in London today. I'm actually missing the airport. I read somewhere that people who miss flying can have airline food delivered to their houses.

I was a bit concerned that my *Sunday Independent* piece saying we should all wear masks might give rise to an attack from the usual nasty suspects. In the UK and USA the anti-mask lobby is very vocal and strong. I don't get it. It's like making taking chemotherapy for cancer a political issue. But to my delight my piece was mentioned on *Brendan O'Connor* and there was agreement that we should start wearing masks! Maybe it will have a positive effect.

Went for a walk with Marg. Down to the sea then along the Metals to Dalkey. Home and had dinner and

then we watched *The Elephant Man*. It resonated hugely.
I wonder why?

MONDAY 27 APRIL

Couldn't wait to get back to the lab. Again, I was the
only one wearing a mask and I'm now glaring at people.
It's hard to glare just with your eyes. Unless I can do a
Christopher Lee in *Dracula*.

Taxi into *Claire Byrne Live*. She took an antibody test
as she had had COVID-19. Three separate tests turned
up blank. These tests detect the level of antibodies in
a blood sample. They are different to the PCR test,
which is a highly accurate lab-based test that detects the
RNA from the virus. I told Claire measuring antibodies
was not as straightforward as it seems. Antibodies are
likely to wane after you're infected. If they didn't your
blood would be churning with antibodies from all
the infections you've ever had, which would be very
inefficient. Instead, you will have more memory B cells.
They will then be primed to make loads of antibodies
when you're reinfected. That's how the immune system
is supposed to work. Me explaining immunology to
the Irish public on a Monday evening. Who would've
thought?

It was a really good chat. Better than wiping down
a shopping-trolley handle.

TUESDAY 28 APRIL

Spoke about interferons with Pat. Interferons are the fire-blanket that can put out the COVID-19 fire. They are important proteins made by the immune system that stop viruses from making copies of themselves – a natural powerful antiviral process in our bodies.

Mark Lawler, an old classmate of mine from Trinity, was on to me about how he had a report coming out on cancer patients not being treated because of the focus on COVID-19. Mark leads a big oncology group in Queen's. I spoke about this issue on Pat's show in order to raise awareness, as it's concerning.

Then into the lab for a huge Zoom call with Sitryx and Eli Lilly to discuss progress – the first one with all the scientists. We went through each of the projects. It was really good.

WEDNESDAY 29 APRIL

Back on with Pat again, this time talking about bats, and how they are reportedly the original source of infection. Someone sent in a question, criticising us and saying we were picking on bats. The *Independent* came into the lab and filmed me making a cotton mask with staples and elastic bands. I took a selfie with my mask on outside Tesco on Pearse St and put it up on Twitter. Putting a mask where my mouth is.

Some interesting science today, too. The antiviral drug remdesivir gave a good result in a clinical trial. It decreased average time in hospital. Sent a piece on this to the *Sunday Independent*. My editor said it might go on the front page! I wrote: 'Before this trial, there was nothing special you could give patients that would have an impact on COVID-19. Now there is. It's called remdesivir. It might be the very thing we've been looking for.' I thought it was good to sound a positive note, as it is indeed the first drug treatment shown to have any effect on COVID-19. Even if it only decreases time in hospital rather than stopping disease, that will be a good thing.

THURSDAY 30 APRIL

The Irish Times came into the lab to film today. Second filming in two days. I wonder if this will be a recurring thing. A film crew used to come in maybe once every six months when we had an interesting discovery. Now they're coming in every week.

Next, *Prime Time* came in and filmed. The total number of cases in Ireland now exceeds 20,000, and there have been 1,232 deaths. I went down to the lobby to get the Trinity Biomedical Sciences Institute backdrop. Paul was the reporter. We spoke about masks.

Rounded the day off with Ivan on *The Tonight Show*. I pressed home the wearing of masks – said it would stop the embers spreading. Held up mine to show everyone at every chance I got. Hope I don't become like a broken record.

April ends with a lot of science. First vaccine efforts. First antiviral. And me banging on and on and on about wearing a mask. And my wonderful lab, back in the saddle, working on COVID-19.

MAY 2020

FRIDAY 1 MAY

Almost balmy today. Waiting for the Dart in Sandycove, the sun warmed my face. The metal seat wasn't wet for once! Simple pleasures. Too much media over the next few days. Will lay off for next week, apart from Newstalk of course. The media needs diversity when it comes to COVID-19, and indeed every topic. If I say no, it gives someone else a chance. And when I drone on, people stop listening, which defeats the whole purpose.

Every time I'm on TV I get negative comments thrown at me on Twitter. And occasionally an email. It doesn't really bother me, but if I'm in the wrong kind of mood I can dwell a little. But it's easy enough to think of something else and remind myself why I'm doing it.

Saying we should wear masks really annoys some people, as does whenever I mention vaccines. Or anything to do with pharmaceutical companies. And, of course, there's always general nastiness. I'm getting a real insight into what it must be like being a politician.

Lab meeting. The four of us in the lab and the rest by Zoom. I told them all to have a drink in their hands and to tell a joke. This fell flat. As Tommy Tiernan said, when asked if he would do a stand-up routine by Zoom, 'I'd rather shout at fucking trucks!' As ever, on the money is our Tommy.

A lovely thing happened. Got sent some great masks by 13-year-old Lauren from Barna, Co. Galway. She wrote a letter saying that she was inspired by my advice to wear masks and started making them. She goes to her local post office every day to post them to various well-known people. One of them has chemical symbols on it. She wrote: 'I will keep making the masks as long as my sewing machine works and Mum keeps giving me money to buy the materials.' This, I have to say, brought a genuine tear to my eye.

The Taoiseach yet again announced another extension of the restrictions, to 11 May. There's also going to be a five-stage plan to allow for a staged loosening of restrictions. This could get complicated.

The FDA today granted emergency-use authorisation

for remdesivir. We have the first antiviral drug approved for COVID-19. The data looks good, but not great. I can't help but think of Tamiflu, the antiviral used against flu. That doesn't really work great either... Remdesivir works by blocking the RNA polymerase in the virus, which is needed for the virus to make copies of itself. It's been shown to decrease time in hospital, which is a good thing. But it doesn't affect the death rate, presumably because it doesn't work on severe disease. If the virus is replicating like mad and really irritating the lungs it looks like it's too late for remdesivir. Therapies are of course the other option, especially if it's going to take a while to get efficacious vaccines. But if the disease really gets a foothold, it's a bit like closing the stable door after the horse has bolted. Vaccines are best because they prevent disease, and to use another well-worn phrase, prevention is always better than cure. Still, let's see what other therapies are discovered. I'll bet there will be a few.

SUNDAY 3 MAY

Another red-letter day! Why? Because it's Desiree's 90th birthday! We all love her so much. We all went to the nursing home and they let us put up bunting and balloons just outside the back patio. She was delighted to see the family: Johnny, Mary, Esme, Ciaran, Fiona, Orla, and us – Marg, me, Stevie and Sam. They wheeled her out and sat her in a comfy armchair. We could wave at her and talk to her over the low wall. They had her dressed up beautifully in a lovely purple dress and elegant grey cardigan, and her hair combed all nice. We raised glasses of champagne and the staff brought out a lovely big cake. None of us could go near her or give her a hug. We were all longing to do that. She smiled all the way through it. She has such a lovely smile and she's such a lovely person. I couldn't help wonder: what the hell has happened? COVID-19 is yet again a malign presence in our lives, stopping us from doing what we really want with our loved ones. We're all wondering that all the time now. We had been planning a huge celebration for her 90th. She waved at us, still smiling, as they brought her back inside and we all went home.

Phoned Helen in Brighton. It's also her birthday and she's 64. We recorded 'When I'm Sixty-Four' for her. Sent it to her by WhatsApp. Me on guitar. Sam on piano, Marg on clarinet and Stevie on percussion. I think we were all in different keys. She loved it! Vera, Chuck and Dave.

Stevie is studying for his exams. I helped him a bit with the ATP synthase. Strange to think he's now 23, doing finals. I thought back to how I helped him when he was in school, with the hard sums. That seems like years ago now. I feel a pang of bittersweet memories.

Why is that? Passing of time hangs heavy this evening, what with Desiree turning 90 and Helen 64. Time just keeps on slipping, slipping into the future. Played that song by Steve Miller on Spotify and, as ever, the magic of music lifts me up to the higher ground.

MONDAY 4 MAY

Bank Holiday Monday. Bit of a nothing day. Made me think of dimanchophobia, a psychological condition that means you hate Sundays. Mind you, today is Monday, but it's like a Sunday. Every day is like Sunday now.

It turns out dimanchophobia is quite common. It strikes people in the afternoon. I remember Jimmy Porter, the lead character in John Osborne's *Look Back in Anger*, as he watches his wife iron his shirts on a Sunday afternoon and says, 'A few more hours and another week gone.' Maybe if he ironed his own shirts he wouldn't be bored.

It might happen because our brains ramp down at the weekend and then have to ramp up again for the working week ahead, and that gives a sense of dread. Funny how we respond to different days of the week. Monday blues, but then as the week progresses, Friday comes around and people's moods pick up. The Cure got a song out of it. We are slaves to our moods. These days I hardly know what day of the week it is, but today I feel it.

TUESDAY 5 MAY

Got some sad news today. Bruno Orsi has died. He was one of my lecturers when I was a student in Trinity in the '80s. He was a tremendous teacher and superb enzymologist. I would count him as one of my key influences. He gave me great advice when I asked him about what I should do after I got finals. He said, 'You're a pretty easy-going guy, so make sure you choose who you work with carefully.' He was warning me off working with some people in Biochemistry in Trinity who were perhaps too aggressive, and I took his advice. When I went for the interview in London, I hit it off with Graham Lewis, who would become my supervisor, and remembered what Bruno had said. Just shows you how a little conversation with a young person can go a long way. I try to remember that when a student comes to see me for advice.

Big lab meeting today on COVID-19, socially distanced as usual. One of the team, Sarah, measured the cytokine IL-18 in patient samples from St James's hospital. This is an important immune protein that can help fight the virus, but if there's too much it can drive harmful inflammation. She found it to be much higher in patients with severe disease. It might be an important marker that tells us about severe disease, or it might be worth blocking. It adds weight to our ideas on my favourite protein NLRP3 as being important in COVID-19, because NLRP3 drives IL-18. This could be a great target to fire our NLRP3 blockers at. It's just exciting to be measuring things in patients, as it brings us closer to what's going wrong in the disease.

Data from Tristram, another member of the team, could be very exciting. He is working on coagulation, and from what other scientists have shown, COVID-19 really affects the blood, causing it to clot in the lungs and other tissues. This seems to be a huge cause of illness and even death. On Newstalk I'd gone into how certain blood groups might somehow protect against COVID-19, although it's not at all clear how that might work. People who are blood group A or B seem to have a slightly decreased risk of death from COVID-19. All this information points to COVID-19 as being in part a disease of the blood. Tristram has some initial data showing that itaconate can block tissue factor, which is a key driver of clotting. Woohoo!

Call with Inflazome this afternoon. Roche want a bit more on the strategy of our plans for treating patients with various inflammatory diseases, so we worked on that. The risk always is we are giving them free information. What if they turn us down? There has to be trust of course, and what we're giving them is pretty obvious, but still I wonder sometimes.

Finished the day off with a live interview on Canadian TV with an Irish reporter called Redmond. Strange to think I was beamed into homes in Canada. No risk of haters coming at me, but then again you never know! He asked me to comment on the UK having the highest death toll in Europe, at over 30,000. A lot of lives lost to this virus.

Sam had his first online exam today. He said it went well. Yet again, extraordinary times with students everywhere doing online exams.

WEDNESDAY 6 MAY

Pat and I covered the symptoms of COVID-19. Barrage of texts came in. Such an appetite. Eimear, the producer, said she's seen nothing like it. I also tweeted a picture of me wearing my mask on the Dart.

Read an interesting piece in *Nature* today on the history of coronaviruses. It said in 1912 a veterinarian spotted strange symptoms in a cat: fever and a huge swollen belly. Looks like this is the first report on what coronavirus infection can do. In the 1960s, scientists in the US and UK isolated a virus from people with the common cold, and they saw coronavirus for the first time – crown-like structures, hence the name. In the 1970s, lots of different ones were found in different animals; scientists realised that dog coronaviruses could infect cats, and cat coronaviruses could infect pigs. Sound familiar? Up until SARS, scientists thought coronaviruses only caused colds in humans. But then SARS came along and had a mortality rate of 10 per cent. And then MERS was discovered, with a mortality rate of 35 per cent. And now SARS-CoV2. A rogues' gallery of coronaviruses. SARS-CoV2 is a big virus compared to others – 125 nanometres in diameter, which in the world of viruses is big. It has 30,000 nucleotides in its genome. Recipe of a killer. And it can correct mistakes as it replicates. This means that very dangerous variants might not emerge, but we don't know. We are bound to see them if it runs riot, as every time it divides there is a chance of a new variant. But flu mutates at around four times the rate of SARS-CoV2, and HIV, 100 times. So it's not a great mutator, which will hopefully be a good thing.

The report said that coronaviruses probably first arose at least 10,000 years ago, but possibly up to 300 million years ago. Some range! The four that cause the common cold are called OC43 and HKU1, which jumped from rodents into humans, and 229E and NL63, which jumped from bats into humans, just like MERS, SARS-CoV1 and SARS-CoV2. One study is concluding that SARS-CoV2 might have emerged 140 years ago, living in bats or maybe even pangolins before jumping into us late last year. What a journey. It changed in either the bat or the pangolin, making it infectious to us. That change is in 4 per cent of its genome, as it's 96 per cent identical to the nearest bat relative. A closer relative may well be found, maybe in the pangolin. As ever, unknowns abound, as we wait for the research to be done to enlighten us.

What is clear is SARS-CoV2 has properties that can be seen in both the SARS/MERS branch of the family and in the other branch, the four coronaviruses that cause colds. And that, like SARS, it can infect epithelial cells in the lungs. But like the others it can also infect epithelial cells in the upper airways. This causes the trouble: if it stays in the upper airways it can spread. In the wrong person at the wrong time it can penetrate the lungs and cause severe disease. One estimate is that if the person beside you releases 100 viral particles, some might reach your lungs. If your immune system doesn't stop them for whatever reason, it reaches the lungs, where it can be as deadly at SARS itself. Even worse, SARS-CoV2 has been shown to infect other cells too – in the heart, intestines, sperm, eyes and brain. This is why some have a whole range of symptoms.

One good thing I read today was that the related virus OC43, which causes a cold, was once more deadly. It mutated and became more benign. The same may happen with SARS-CoV2.

FRIDAY 8 MAY

Friday … but you wouldn't know it.

Jennifer O'Connell visited my lab for a piece she's writing for *The Irish Times*. It was really great. One of the best interviews I've done, I think. Yet again, superb journalism in Ireland. I told her I'd written some of the book in south San Francisco and she told me all about her time there. She said she remembered going to a great food market in San Carlos (where I stayed). We had a really good discussion about COVID-19 and all the current issues. It was great meeting her, as I love her Saturday columns in *The Irish Times*. She told me to ignore all the haters on Twitter – she has plenty herself! She said how her mother was a big fan, which was nice. It was just lovely to spend a couple of hours with her.

After that, I gave a talk at a conference on immunometabolism – so strange, being able to jump from one thing to another in my office. Jeff Rathmell from Vanderbilt University in Nashville organised it. I really wanted to go, given all the musical associations of that city. All so different from doing it from a podium in front of a big audience. Wonder if the talk will have the same impact?

Got me thinking about how science has changed. No more face-to-face conferences. Hard to know if this will slow progress. It certainly makes it all less fun, and, for me, out of fun comes new ideas and insights. Talk went well though. Lots of questions. Maybe I'm getting used to it.

Zoom drink later with my old college friends John, Peadar and Bren. It was good, but jeez, no substitute for going to the pub and knocking back a few pints and slagging the holes off each other.

Really relaxed tonight. Giving the big talk was worthwhile. 'Nights in White Satin' just came on the radio, and it made my soul swoon. If your head is in the right place a piece of music can really register and transport you somewhere elsc. Truly makes us human.

SUNDAY 10 MAY

A stormy day. High winds whooshing through the leaves on all the trees. Love that. This morning Brendan O'Connor and I spoke about possible treatments, including remdesivir, and how children are largely being spared the virus, which is a huge relief. It's still not fully clear what the death rate from the virus is. More than flu, which is around 0.1 per cent of cases. It might be as high as 2.5 per cent but this will depend on age. People who keep saying it's just another flu are plain wrong. I told him about the frenzy of effort to find a vaccine, which will be the ultimate weapon, but tried to manage expectations. Really, we just don't know if we'll get a vaccine.

Went to Dunnes in Cornelscourt with Marg. Took half an hour to get in with the queueing but it was worth it. It's a sad and sorry state of affairs when I get my kicks from queueing for the supermarket shop.

MONDAY 11 MAY

Sam's birthday today, yay! Made him his favourite duck salad. His mates came over and they drank cans in the garden. We made the most of it for him.

I'm a bit sick of COVID-19, it has to be said, but took my mind off it all by watching a documentary about Dana. I'm old enough to remember her winning the Eurovision with 'All Kinds of Everything'. I was five at the time and our teacher got us to learn the words. It still lifts me: '*Sailors and fishermen, things of the sea …*' Good God, they were innocent times.

And for the first time in its history there will be no Eurovision song contest this year. It was to be the 65th contest. I can see in years to come when they look back on the Eurovision, this will be mentioned as the one and only time it wasn't held – because of a virus.

TUESDAY 12 MAY

Had to sign lots of share certificates for the staff of Inflazome in my capacity as director. People are buying shares! They would be stupid not to.

Did my first set of exam-script corrections for the final exams in immunology. This was a new experience! Software revealed whether students had plagiarised, and we were instructed not to mark those parts. Impressively, not many had – they must have heeded the warning.

Slipped out to buy some presents for Marg's birthday. Almost nowhere open. It's just really weird to be walking up Pearse St with hardly any cars on the road. And then Grafton St with nobody about. Alex told me that Tesco in the Jervis Centre was open and had party stuff, so I headed over the Liffey. Got streamers and a banner. Then went to a Dealz that I noticed was open – essential retail, obviously. Jackpot! Got lots of stuff for the garden: flower seeds, a birdhouse, a face to stick on a tree and a hanging metal frog that clangs. I spent a total of €20, but hey, it's the thought that counts. Wrapped it all up in 'Happy Birthday' paper – result! It's the little things we can still do, like wrapping birthday presents, that still give us pleasure.

RTÉ came in to the lab and interviewed me about face masks. It was on the six o'clock and nine o'clock bulletins, just after Tony Connelly reporting on Europe. Surreal or what? When me and Tony were both broke and living in London in the 1980s, we never imagined in a million years that we'd be back-to-back on the news, now did we?

WEDNESDAY 13 MAY

I noticed a slight increase in people wearing masks on the Dart, but still only 20 per cent or so. I obsessively counted people in the carriages in Pearse Street as I walked down the platform. We have to do much better.

THURSDAY 14 MAY

Marg's birthday! A day filled with light, sunny, blue, white. Started with breakfast – made her eggs Benedict and brought her a tray in bed. Gave her the gifts. Think she liked them. Then I made a chocolate cake. Turned out good, and I covered it in icing. We had a wee party in the front garden. Sam cooked crab claws, and I cooked trout. We'd planned a big party, but that was yet another COVID casualty.

FRIDAY 15 MAY

Did a journal club with the lab. This is a standard thing in labs. Someone picks a recent scientific paper that they like and goes over it with their colleagues. We did three. Not much data at the moment because the lab is only getting going again. Z picked a famous paper from the annals – the discovery of how Toll-like receptor 4 senses bacteria. The foundation of what we work on, really. And because we're obsessed with COVID-19, we discussed how Toll-like receptors 3 and 7 are sensing that virus. Toll-like receptors are front-line sensors of infection, and were a huge breakthrough 20 years ago when they were discovered. And like so many other parts of the immune system, they have relevance to COVID-19.

Tony Holohan announced tonight that seven children had presented complications from COVID-19. It looks a bit like Kawasaki Syndrome, a multi-organ inflammatory syndrome. Coincidentally, Moshe Arditi, in Cedars-Sinai Hospital in Los Angeles, contacted me. He is a collaborator of mine and a renowned paediatrician who works on Kawasaki Syndrome. This is a horrible disease which afflicts children under five. Their blood vessels become inflamed all over their bodies. It's very painful but thankfully it can be treated with immunoglobulin. Its mercifully rare. Moshe is submitting a big paper on this aspect of COVID-19. He said it looks nasty – not quite like anything he's seen before. Another worry for us all.

In better news, the government has announced that easing of restrictions will begin on 18 May with the reopening of tourist destinations such as the Botanic Gardens and Trim Castle. Seems sensible enough.

SUNDAY 17 MAY
Thinking about my *Sunday Independent* piece, entitled

'A nation holds its breath' (for once they went with my title!). Important to warn about how, with many countries (including Ireland) opening up again, we risk a second surge. And there is recent evidence of how the virus spreads in closed, crowded spaces. I wrote: 'This is literally a matter of life and death for governments everywhere. Get it wrong, and people die. You can't do much about the testing, but you can certainly play your part by following the rules. Wash your hands, maintain social distancing, keep surfaces clean, wear a mask in shops and on public transport. Be unrelenting, because you know what, SARS-CoV2, the virus that causes COVID-19, certainly is.'

All we can do is hope Ireland, and indeed the rest of Europe, doesn't go into a second surge. It has to be said the chances of that are high, because people are bound to get together in large numbers. We just have to hope it isn't too big a surge.

Went to see Desiree in the nursing home. She was allowed outside so we sat in the sun, but it got too breezy for her. Walked home via the People's Park in Dún Laoghaire and yet again, two people shouted at me – both in a good way!

MONDAY 18 MAY

The glorious weather continues. Lucky, I guess. Imagine

COVID-19 in bad weather. What will the winter be like?

TUESDAY 19 MAY

Two calls today about Inflazome and questions about the diligence process. I handled the science ones, but there were loads on the safety studies and the Phase 1 data. That's the phase that checks that the drugs are safe in people. Phase 2 then is the first clinical trial against a disease, usually involving fewer than one hundred people, and then Phase 3 is a much bigger trial. It's all done step by step, to ensure safety and also to decrease the overall cost. No point in doing a huge Phase 3 trial unless the drug has proven itself in a smaller Phase 2 trail.

Also had a board call with Sitryx. They are doing great too. They are very interested in our itaconate work. I told them of the data we have of it blocking SARS-CoV2 growth. They said to keep them posted as they could really help us make it all real for patients.

Had lunch in St Stephen's Green. A woman came over and said, 'Thank you for keeping me sane.' I wondered what my dad would make of all this. He spent years working in the Green with his deck-chair business, so I often think of him when I walk through it.

Rounded the day off recording three messages for schools. Sent them on WhatsApp. I'm getting lots of requests from teachers to record messages for the graduation ceremonies for sixth-years. It's just so tough for them. They are missing out on a key rite of passage in their lives. This bloody miserable virus.

One of the requests was from Pat Gregory, principal of my old school, Pres Bray. He was my old geography teacher and it's his last year there. He said if Matt Damon was good enough for Loreto Dalkey, then Luke O'Neill was good enough for Pres Bray. I was absolutely delighted to do it. I told them about the last day of the Leaving Cert way back in 1981. Me and my classmate Hugh Roche went down the seafront, and I said 'Let's jump in the sea!' so we did. In our uniforms. It felt great and we never forgot it. I said, 'Go for it, lads!'

WEDNESDAY 20 MAY

Recent evidence indicates that if you've had a cold, it might protect you a bit from COVID-19. The viruses are similar, so you might still have antibodies and T-cells knocking around your body that could afford some protection – almost like the cold is acting as a vaccine. Discussed this on Pat's show, and he made the good point that this might protect teachers, who will have had more colds. Let's see if it's true. We also spoke about 'biomarkers', which can be used to predict if someone might be at risk of severe disease. A bit like the IL-18 we measured.

THURSDAY 21 MAY

Had an important call with our Dutch collaborators. Experiments are running there. They also put us in touch with a lab in Belgium who can do in-vivo work on hamsters infected with SARS-CoV2. Hamsters are used because the features of the disease resemble that in humans, so important information can be obtained. Excited about this, as it will be the first time we test itaconate in infection.

Yet another interesting report today. Some people shed far more virus than others and so are super-spreaders. Some breathe out more viral particles just by talking. Singing definitely releases a lot more, hence the link to choirs. All the studies point to the same thing: enclosed spaces, poor ventilation, crowds, people shouting and singing. Add alcohol to the mix and you have an Irish pub. But also a meat-packing plant: the virus spreads there because of close contact between people, a lot of shouting (they are noisy places) and the fact that some of the workers live together in cramped accommodation. Even Zumba classes have been shown to be sources of infection, whereas Pilates classes haven't. Zumba is a high-impact exercise, and must lead to lots of virus coming out of people. Such great science going on.

FRIDAY 22 MAY

Got a paper to referee showing similar data to us –
itaconate blocking SARS-CoV2. So we've been scooped
already! Bah! Still, it shows what we have could be
correct, and there's always room for more research. Had
to recuse myself from refereeing it because of the conflict
of interest. The refereeing process is a very important
part of science. Experts are asked to closely read another
scientist's work to make sure everything is in order, and
that what they are concluding is backed up by their data.
It's an important task as it ensures that what is finally
published is likely to be correct.

Attended an online ceremony to induct the new
members of the Royal Irish Academy. This is normally
a lovely day, as it's an honour to be made a member. I'd
nominated Kate Fitzgerald, former PhD student, now
famous immunologist, and she was inducted. I joined
the Zoom, but missed her induction by five minutes!
We'll mark it in some other way. Good on you, Kate,
your home county of Waterford should be proud!

SUNDAY 24 MAY

Today we covered Stevie in eggs and flour. He told us
that it is the tradition in Edinburgh when finals are over.
Friends turn up outside the exam hall and pounce. So,
we called him out into the garden and *splat!* Not quite

the same as the real thing, but yet again, with COVID-19 we have to improvise.

Had a long chat with Paul Moynagh on the phone. He was in my lab years ago and now he's head of the department of biology in NUI Maynooth. He's pressing for much more frequent testing – and he's right – so that we can hunt down the virus and stop it from spreading. He was one of the first people to be in my lab all those years ago. And now look at him! He's a superb scientist, so why won't people listen to him?

Watched two episodes of *Better Call Saul* with Sam. We love that show. The glacial pace of the drama seems all the more appropriate these days.

MONDAY 25 MAY

Covered super-spreaders with Pat this morning. More and more evidence that spreading is happening in enclosed spaces with poor ventilation, and it's a single person infecting lots of people. Got to keep reminding people to watch out.

Today is the first day we've had zero deaths since the pandemic began. A moment to savour, yes, but we need to be vigilant.

Read an interesting piece today. A report has been released showing that traffic has increased by 30 per cent on the M50 motorway compared to the levels at the beginning of lockdown. The reopening continues. There was also a study by the Dublin Institute of Advanced Studies showing that seismic activity had increased. No, not earthquakes; they can actually measure Ireland move, as people and cars move around. Brings whole new meaning to the phrase 'Did the Earth move?'

Horrible pictures on CNN tonight: George Floyd being choked by a police officer who was kneeling on his neck. Life's brutality goes on in spite of COVID-19.

TUESDAY 26 MAY

Prime Time came into the lab to film a piece about masks. The message is slowly getting through, I think. Even more people on the Dart wearing them. No one was wearing them in Tesco this morning, when I went in for my usual coffee. Not even the staff. FFS.

WEDNESDAY 27 MAY

I went around to Brian McManus's garden for a drink. He still has a keg of Guinness. In answer to a question in a recent interview with the *Irish Independent* on when I'd last cried, I'd said it was when he'd handed me my first pint of Guinness in weeks. (The marketing manager of Diageo sent us an invite to come to the Guinness Storehouse after the restrictions are lifted!) It was a lovely summer's evening. A warm stillness in the air. The three pints made the air shimmer even more.

FRIDAY 29 MAY

Spent the day correcting immunology and biochemistry finals. I was on the lookout for evidence of independent thinking. If we can lift students who might get a third-class degree to a 2:2, and also not let down the ones who are outstanding, then we're doing a good job. It's a tricky balancing act. The student who gets a 2:2 when they were expecting a third is as thrilled as the one who gets a first and wasn't expecting it. This can carry them forward in their lives. But we can't let down the truly remarkable students either. Press them. Challenge them. Set them loose on the world with pride too, because they've also been stretched.

For a break I went up to Dunnes to buy some T-shirts. Yet again, strange to be walking up an empty Grafton Street on a Friday afternoon. I went to the entrance of the St Stephen's Green shopping centre and the security guard looked wryly at me and pointed to the long queue I had missed. I asked him if it was moving fast. And he said, 'How the fuck would I know?' Can't beat the wit of Dublin security men. I chanced it and it took about 30 minutes to get in. This is where iPhones come in, as I could work away on mine. Once in, I made for the T-shirts. A woman came over to me and said, 'You're Luke O'Neill. I just want to say a big thank you. You've been keeping me going.' So kind of her. I still can't get used to being recognised in public.

Back to the lab for a bit, then drinks in Merrion Square with everyone. I had a couple of glasses sitting on the grass in the evening sun. God, the weather has been great. Relief for all the people in lockdown. I wonder what it will be like in the dark winter if the virus is still rampant? Cross that bridge when we come to it.

Went home and watched *Chariots of Fire* with Stevie and Marg. I wonder will Stevie get to run around a quad? Go on my son!

SATURDAY 30 MAY

And so May ends. The dullness of this pandemic only relieved by the good weather. We seem to have the virus under control. So much science now to draw on – avoid the three Cs (close contact, crowds and closed spaces) to stop spread. Dare we hope that there's a way out?

JUNE 2020

MONDAY 1 JUNE

Well now, if there was no COVID-19, The Metabollix would be on our way to Singapore. I was due to speak at a special immunology meeting, and they said bring the band to play at the conference dinner. I have a voucher for the flights, for which I paid up front. I bet I'll never use them.

We went over to our friends Mai and Martin. Had a nice meal in the garden, all consistent with the current regulations, which allow two households to meet in one household's gardens. And something incredible happened: they ordered pints of Guinness from their local pub, which were delivered to the garden. Again, who would've thought such a thing could happen?

TUESDAY 2 JUNE

The government has released an analysis of the COVID death rates in Ireland. Five per cent ended up in hospital, and of those 1 in 3 were admitted to the ICU. The fatality rate was 0.6–1.4 per cent. This is so much more serious than flu.

On *The Tonight Show* with Ivan Yates. Michael Levitt came on to say the pandemic would soon be over in Ireland and indeed all over the world. Ivan asked me what I thought. I couldn't restrain myself. I said 'Rubbish!' Let's see what the future numbers tell us.

WEDNESDAY 3 JUNE

Marg drove me to Cork today. I got a letter giving me permission, as it was for an interview on COVID-19 with Maura and Dáithí on *The Today Show*. While we were down there, we picked our motorboat up from Billy Morrisey's shed. Lucky to be able to do that at the same time. We towed it back to Dublin, launched it and put it back on the mooring. We'll have it now for the summer. Can't think of anywhere safer to be than out on the sea, wind blowing. Maybe I'll just sail away.

THURSDAY 4 JUNE

A huge session with Pat today. Discussed how sampling sewage might reveal an outbreak before it gets a foothold.

This is because the virus can end up in sewage and can be sampled at sewage plants using the PCR test. If it goes up, it means there's an outbreak happening and lots of testing of people can then happen to limit the spread. We also discussed how men are more badly affected than women, with more mortality. Not clear why. Perhaps men have more ACE2 (the protein that SARS-CoV2 uses to get inside cells) in their lungs because testosterone boosts its expression? This means there are more receptors for the viral spike to stick into. Yet bald men seem to get less ill when infected. Female-led countries seem to be doing so much better (for example, New Zealand). In the US, COVID-19 hits lower-paid workers the most, revealing the divides in society. Huge amount of information building up, which I am trying to keep track of.

Word is Europe is opening up to allow people to travel for their holidays. Surely this is a mistake? Without proper test-trace-isolate systems that is a recipe for disaster.

The CEO of Driving Tests Ireland called and we discussed how to make driving tests safe. I said the usual: masks, good ventilation. He seemed reassured.

Wrote piece for the *Sunday Independent* on the battle of science versus economy. Made the case, as others have, that they are inextricably linked. Without getting the virus under control, the economy won't come back. Not least because of consumer confidence.

FRIDAY 5 JUNE

Today was a much-needed 'lazy' day. Lazy because of no media work. Just helping the lab in various ways.

The Taoiseach announced that we're moving from 'Stay home' to 'Stay local'. We're about to enter what is being called 'Phase 2 plus'. Hard to keep track of all this. One thing is clear: no one can leave Ireland unless it's deemed essential. When will I be able to travel again? I really, really miss it.

SUNDAY 7 JUNE

Another day of dimanchophobia. However, went to visit Desiree. This time they let me in to see her in a special room. We sat in seats 3 metres apart. They said we had 30 minutes. There was a clock on the wall behind me, and she kept saying how many minutes were left. We had a laugh over it. She seems happy enough. A Buddhist, living in the moment because her short-term memory has gone. Not the worst way to be, I suppose.

MONDAY 8 JUNE

AstraZeneca said their vaccine trial was well advanced. They must be confident, as they are making 2 billion doses. They've said they won't be making a profit from it, which is curious indeed. Interesting that a big pharma like AstraZeneca have signed up for that.

Got an excellent email from the journal *Cell Metabolism*. The referee reports on Alex's paper are in and they are reasonable! This is unusual, as often they ask for a lot more experiments. But it looks like in this time of COVID-19 they are being a bit more lenient as they know labs are working at full capacity. The experiments they are asking for are straightforward. Alex is delighted, as indeed am I. I went to tell him but he was in the toilet, so I waited outside, and told him as he came out. He beamed away! Makes all the effort worthwhile, as it's so important for his CV, and our data will be out there for all to see.

Got home at 8 p.m. Sam was working on 'Candemic', his Instagram beer-review page. He's also making T-shirts and masks with 'Candemic' on them. I said he should contact various brewers and ask for free samples. Wicklow Wolf and O'Hara's would be good. He may well turn into an entrepreneur!

TUESDAY 9 JUNE

Another big call today about Inflazome. Roche are still interested (phew!). They will get back to us soon. It would be terrible to get so far and then for it all to unravel. Still, it's happened to me before with Opsona (my first start-up biotech company, which was almost bought at one point until the deal fell through at the last minute) so I must remain strong.

WEDNESDAY 10 JUNE

Hugely productive day. Funny how sometimes you get things done like they've never been done, to quote good old John Lennon, in the song 'Hold On' from *The Plastic Ono Band*. A remarkable album in its rawness.

Ciana, another graduate student in the lab, is making progress on itaconate acting a bit like a steroid. This could be good for our COVID-19 project, as the steroid dexamethasone continues to show promise as a therapy to treat severe patients. It reduces the death rate by about 20 per cent. What if itaconate could do the same or better – wouldn't that be good?

On *Drivetime* talking about masks. Again. And then gave 1,300 GPs a COVID-19 update via Zoom. Never have I spoken to so many people in so few months.

THURSDAY 11 JUNE

Sometimes I wonder what I'd do without my regular Pat ritual. Today I informed listeners that data has changed and surfaces aren't quite as dangerous as previously thought, though of course we still need to keep surfaces clean. And another thing: cows could be a great source of antibodies to treat COVID-19. They make twice as much antibody per millilitre of blood as humans – and of course, being large animals, they have a *lot* of blood. The cows might save us! John from *Claire Byrne Live* rang me after and wondered if we could do something on that … get a cow into the studio? Or a life-size model? Great, I said!

Made four more videos for schools, including one for Lucan Educate Together. Yet again, blown away by the commitment of our teachers – we are so lucky with our education system.

Bit of a hoo-ha in the Dáil today. NPHET are still being unclear about masks, saying the evidence wasn't 'great'. I tweeted on how outrageous that was. Meeting Damien Nee, who was on the group advising on masks and other areas, tomorrow.

FRIDAY 12 JUNE
Damien came in with Duncan Smith, Labour TD. Duncan had asked the question of NPHET regarding masks. He is strongly of the view that they should be recommended.

Met up in Brian's garden for a few pints. My old pal Cormac Kilty came along. We had great chat, which reminded us of how things used to be.

SUNDAY 14 JUNE

Put the absolute finishing touches to *Never Mind*. Satisfying in the extreme! Added in a few more COVID-19 references. Got several digs into Trump. Was outraged when I saw him say live on TV that he would inject bleach, and then saying that he would just take hydroxychloroquine. 'What have you got to lose?' he said. 'Your life,' said the experts.

Went for a walk tonight and felt like I was actually walking on air. Not sure what it is about me, but I find I can fully relax after I've done some worthwhile work. And because I was raised a Catholic I can't even blame the Protestant work ethic for that! I have a sense that the coming week is going to be mega.

MONDAY 15 JUNE

Today was all about *Claire Byrne Live*. We moved some of my lab into the studio. We brought a PCR machine, an ELISA plate reader and various bits and bobs. And then Eva from my lab demonstrated how the ELISA machine works. She explained that the ELISA is a very sensitive way to detect antibodies. We discussed antibody tests and I pointed at the PCR machine and said that's where the virus can be measured, whereas the ELISA machine is where antibodies can be measured. They had made a great prop for us: the SARS-CoV2 virus, which is like a small football, and an antibody molecule. I'll bet they'll come in handy for future demos. Eva did a great job – it's not easy pipetting live on TV.

The FDA revoked its emergency-use authorisation for hydroxychloroquine, saying the various trials that have been done have failed. This is how science should work: idea, test, get data, come to a conclusion, repeat. Bottom line from all that is hydroxychloroquine provides no benefit in COVID-19. Are you listening, Mr Trump?

TUESDAY 16 JUNE

Another important day in the fight on COVID-19. They are coming thick and fast now. The data on the drug dexamethasone was published. It has been shown to protect people with severe COVID-19, decreasing deaths in one third of those patients who would otherwise have died. I had worked on this during my PhD as an anti-inflammatory, so I took a picture of the graph on it from my PhD thesis and sent it to Pat. This is good news. Doctors now have an anti-inflammatory to help people. Looks much better than remdesivir.

WEDNESDAY 17 JUNE

My birthday! Fifty-six years of age now. Ah sure, what of it, eh? Another year goes by, and what a terribly strange one. Boys gave me pressies before I headed into work. A nice camera! In the lab, Cait brought in a cake for me. She always does that – so kind of her. And my favourite, Victoria sponge! Stevie then came in by boat – first journey this season. He came up the Liffey and picked me up at the Convention Centre. Very thoughtful of him. We speeded out to Dún Laoghaire.

Marg bought me a new fancy outdoor fireplace – don't know what to call it, chiminea or something? It looks Scandinavian, but it's Mexican and was made in Donegal. We put some turf in it, lit it and had a few drinks. Feel like I've got a cold coming on. Urgh. Or could it be …

THURSDAY 18 JUNE

Felt fine on waking. Just goes to show – the vagaries of our immune systems. Trying to ram home the message about masks this morning. Pat coined the great phrase, 'My mask protects you, your mask protects me.' We discussed other coronaviruses and I told him about SADS-CoV, which infects pigs and stands for Swine Acute Diarrhoea Syndrome. At least SARS-CoV2 doesn't cause acute diarrhoea.

BBC Radio Northern Ireland, then *The Six O'Clock Show*. Met Eilish O'Carroll, aka Winnie in *Mrs Brown's Boys*. She said she'd found lockdown very tough. She tried cooking, but as she lived on her own she couldn't see the point. 'I'd bake some scones and then eat them all.' Who would have thought that the baking of scones would be an existential threat? Brought home yet again the difficulties people have been going through.

SATURDAY 20 JUNE

Got into Cavistons! Only took 20 minutes. A woman in the queue said to me that the virus was now gone from Ireland, so what's all the fuss about? It felt like a magical wonderland. Bought three types of fish and some smoked duck. My God, we'll have a feast.

SUNDAY 21 JUNE

Father's Day – at least the lads remembered. Brought me breakfast in bed. Went out in the boat with Stevie, around Dalkey Island. Nice sunny day. Lots of boats out. No virus out here.

Watched an interesting show on BBC called *The Lumineers*. Eve Hewson plays the lead role. She is a compelling actress. Set in New Zealand in the 1800s. Unusual and interesting. I'll watch it every week.

MONDAY 22 JUNE

Sweden didn't have a big lockdown and is being held up by anti-lockdowners as an example of what to do.

They have had more deaths than their Scandinavian neighbours, but some say it was a price worth paying, as they might now have herd immunity. There is no evidence for that at all, and it's not ideal to compare countries. Sweden has fewer people on average in each household, which might decrease case numbers. All kinds of variables at play. Limit the variables when you're looking for an answer. I predict Sweden will look back and regret the death rate it has seen.

A man who went into five nightclubs in Japan infected 96 people. One person in a poorly ventilated restaurant infected 10 people out of 89. I want to let people know that they have to take care.

I interviewed Bill Campbell by Skype for the Royal Irish Academy. I asked him why he was so obsessed with parasitic worms. He said it was because they were so ingenious and beautiful. He's 90 and has never lost his Donegal accent, in spite of living in the US for decades. He is such a lovely, modest man. Like so many, he hasn't seen his grandchildren in months. He's a bit sceptical about ivermectin being useful against COVID-19, which has been reported. Winning the Nobel Prize in no way fazed him. He said he and his wife had a tremendous time in Stockholm. His wife's main worry was what dresses to wear to the events. He said a big thrill was going to the White House, and Barack Obama giving him a parasitic heartworm plush toy. A drug Bill had discovered is used to treat heartworm in dogs, so not only has Bill helped prevent blindness in humans from a parasitic worm, but he also saves man's best friend. What a hero.

WEDNESDAY 24 JUNE

Emphasised on Ryan Tubridy this morning that everyone should avoid the three 'Cs': close contact, closed spaces and crowds. This was first used by public health authorities in Japan, and it's an easy thing to remember.

THURSDAY 25 JUNE

Alex is almost ready to resubmit his paper. Went over it and it's looking good, hooray! I am optimistic, but until you get the acceptance from the journal you can never be certain. Today was a glorious day, so I thought, to hell with it. Out in the boat. My old college pals Peadar and Jock came along. We anchored off Dalkey Island. There were lots of boats there. Stevie spotted a huge speedboat coming along, with someone on jet skis alongside. He said it was Conor McGregor – and indeed it was. It felt like the Irish Riviera, at least for one day. Marg went for a swim off the boat with Stevie. Sam jumped in too. We then had some cool beers. No fear of the three Cs when you're at sea, with a lovely warm summer breeze blowing.

At last Simon Harris has announced that face coverings are now mandatory on public transport. A step in the right direction. Hopefully we will become a nation of mask-wearers, just like the Asian countries.

Another important milestone today. China have approved the first vaccine for COVID-19! But there's a trick – it's only for the military and they haven't done a proper Phase 3 trial. In effect, what this means is the army gets the vaccine and the Phase 3 large-scale trial runs in parallel. Unprecedented but hugely important. They must be confident that it will work.

FRIDAY 26 JUNE

Ordered pizza and watched Elbow playing at Glastonbury on TV with Sam and a couple of his mates. We all wondered when we will get back to gigs again.

Meanwhile, society is reopening all over the world. I saw a report that said half the world's population has been in some kind of quarantine. But new deaths and cases are plunging in Europe and Asia. There's a prediction that the worst of the pandemic is behind us. I hope so. But the UK reported its largest-ever fall in GDP.

SATURDAY 27 JUNE

Well, well … Fianna Fáil–Fine Gael have agreed to form a coalition. Tough road ahead for them. I wish them well.

SUNDAY 28 JUNE

Bit of a hangover, it must be said. A few too many beers last night, but jeez, I felt I deserved them. Bill Campbell interview was broadcast by the Royal Irish Academy.

Noticing a fair few attacks on Twitter. To be expected. They seem to fall into four camps: anti-maskers, anti-vaxxers, people accusing me of being in the pay of pharmaceutical companies, and what I call general nastiness. Reminds me of being in the playground at school, and someone shouts abuse at you. Twitter does not lend itself to debate. I think it should only be used for information exchange. Intriguing to watch and to consider the motivations behind the negativity. Fear? Jealousy? Hate? All the usual gamut of human emotions and frailty. If I was younger, it might bother me. But I'm too old to worry too much. And a lot of people send nice stuff too. As Rudyard Kipling said: 'If you meet triumph and disaster, and treat both imposters just the same, you'll be a man.' My dad used to say that but would change the end to 'you'll be a bloody idiot'!

We are now six months into this pandemic. More

than 10 million cases, and over 500,000 deaths. It just keeps climbing and climbing, in spite of these indications that it might be slowing down. And people still keep denying it.

MONDAY 29 JUNE

Another Monday, another session with Pat. Covered lots of things, including special organoids, grown in a lab, that are like lung tissue and can be used to study infection. But I wonder how long can we go on like this. When will we go back to weekly sessions on the latest science, whatever that might be? It's the biggest science story since ... the moon landing? Reminds me of the line in that song from World War One about American soldiers going back to the US: 'How you gonna keep 'em down on the farm, now that they've seen Paree?' In my case: 'How you gonna talk about any other science news, now that you've been going on and on about COVID-19?'

And today was another good day because ... guess what? Hairdressers, barbers, gyms, cinemas and churches have all reopened. Pubs serving food also reopened but not 'wet' pubs, as they are being called. Another new term. Tony Holohan has pointed out that a packet of peanuts does not constitute a meal. He knows the Irish well. The excitement of it! This is Phase 3 of the reopening. Wouldn't it be great if we never have to go back to Phase 2?

TUESDAY 30 JUNE

Twitter bugging me a little still, but more disturbingly, got a nasty email. *Next time it won't just be an email. I will come to your house.* This gave me pause for thought. I'll soon need a bodyguard! What seems to rile people is being told what to do. Don't they realise my only intention is for the greater good? Still, all over social media I see people being attacked all the time. These are the times we live in, I guess.

Gave a talk to the senior management at AIB. AIB have kindly given €2.5 million to our COVID-19 centre, which is just starting up. It means lots of labs in Trinity working together, some in the Trinity Biomedical Sciences Institute and some at the Trinity Translational Medicines Institute up at St James's hospital. Working on everything from testing to vaccines to therapies. We also have funding from Science Foundation Ireland, so we are up and running. Great progress indeed. Lots of questions on where we are at. Also more and more about the vaccine prospects, so I told them about China's progress.

The two biggest things with COVID-19 this month are the gradual reopening (here's hoping) and the

Chinese vaccine being approved. In other news, *Never Mind* is off to print. And the Roche uncertainty. All these parts of my life interlinking and turning like one big machine inside my head – it's no wonder I have to douse it with beer (nature's lubricant) every so often.

JULY 2020

WEDNESDAY 1 JULY

Both S and Z were offered jobs today! S got an academic post in the University of Graz, and Z got a job in a company back home in Warsaw. Great to see them both developing their careers. As it should be, but I'll be sad to see them go. The lab turns over. I think I've probably had six sets of people over the years. Several have gone on to academic posts all over the world – in the USA, Australia, the UK and Ireland. I counted – 18 of them now have their own labs, which is great. And some of them are in the media talking about COVID-19. Paul Moynagh, Annie Curtis, Beth Brint and Clíona Ní Cheallaigh all worked with me. Tell that to the haters!

Gave another Zoom talk tonight at the Royal Irish

Academy. It's so important to have learned societies like the RIA, who host properly moderated discussions and events – not like Twitter or people shouting through megaphones at anti-mask rallies. Shame on those who use social media to vilify others (not talking about myself there, you understand!).

THURSDAY 2 JULY

Another incredible day on this COVID-19 journey. Good Lord, my life is so different now. Intense but also just surreal and 'out there'. On the Dart I counted 6 out of 15 people who were wearing masks. Not good enough, but better than before.

Sometimes I think the show should be called *Pandemic with Pat*. Informed our listeners that all evidence now suggests that COVID-19 jumped from a bat or a pangolin into a human. Like its relatives SARS and MERS, it began in another species, mutated slightly and then jumped into us. The mutation made us acceptable as a host. But because it's a brand-new virus, our immune systems have never seen it before, and so as a population we have a low level of protection. For each person we have to mount a fresh defence against this new pathogen. Younger people, and healthy people without other diseases in general, can mostly fight it off thanks to our adaptive immune system. But for the vulnerable, which mainly means older people, or those with lower immunity, it can wreak havoc.

So will it happen again? It might, but it's impossible to predict. Science doesn't always have the answer, but strives to find it. There are 1,400 different pathogens known to infect humans, of which 20 per cent are viruses. Many have jumped from other species – as much as 70 per cent. HIV jumped from monkeys. H5N1 flu virus jumped from geese. There may be as many as 700,000 other coronaviruses in animals. Scientists are trying to track them all, to stop all this happening again. A 'Global Immunological Observatory' is being set up and even involves collecting bat guano to look for viruses. Brings whole new meaning to the phrase 'bat-shit crazy'.

A huge delegation came to the Biomedical Science Institute in Trinity today. Minister for Health Simon Harris, the Director of Science Foundation Ireland and the CEO of AIB all in my lab. The opening of the COVID-19 Research Centre was announced. My colleagues Kingston Mills and Aideen Long had done a great job at pulling it all together. And now here we are. It's great that TCD and Ireland can play its part in this fight against the biggest public health crisis in 100 years.

Then in the evening I took part in the 'Provost's Salon'. This is a discussion hosted by the Provost, and supporters of Trinity are invited to attend. An amazing list of attendees were on the call tonight. The Earl of Iveagh took part. He has been a great supporter of Trinity over the years, and he asked some very interesting questions. As did the managing director of Microsoft Ireland, Cathriona Hallahan. I spoke for around 20 minutes and then the deluge of questions came in. As with everyone else that I speak to, there is such an appetite for information. The people invited may well help us in various ways, and I look forward to further engagement.

A remarkable day, really.

SATURDAY 4 JULY

A nice, relaxed day. Tony Holohan announced that he is stepping down as Chief Medical Officer to spend time with his family. It's widely known that his wife is suffering from cancer. It's a tough time for him. The whole country feels for him, and there has been huge support. People are pressing for him to be given the Freedom of Dublin. Irish people can be so decent. It's something I am especially proud of.

SUNDAY 5 JULY

Got a spread in the *Sunday Independent* comparing COVID-19 to the movie *Jaws*. Quint is the front-line healthcare workers who risk their lives. The mayor represents the business people, trying to open the beaches to save the summer and yet the shark is still out there. Hooper is Tony Holohan, coming up against business: 'Like to get your name in *National Geographic*, wouldn't you, Hooper?' The article makes a point, I hope.

MONDAY 6 JULY

First signs of the virus mutating. There have been at least 1,300 separate strains coming into the UK. Changes are in the spike protein. There's hope that these won't affect the virus too much but still, we don't know.

I'm currently getting around 15 emails a day with questions and try and answer them all. For example, choir leaders are worried about not being able to get together. Choirs are so important for communities and for people's well-being. And yet this virus likes nothing better than a choir in which to spread. All I can advise them to do is to sing outdoors, keep as much social distance as possible and if at all possible, wear a visor or even a mask. It's so difficult. I would love to tell them all will be well soon.

Others who have contacted me recently are people in rowing clubs, hockey clubs, teachers, mixed martial-arts instructors, worried parents and grandparents and even driving instructors. Here are some of the things I've been asked:

I am desperate to hug my lovely grandchild. What's your advice?

....

Answer: it's OK to have a quick hug outdoors, wearing a mask.

I am a film director, and we are about to start shooting a movie. I want to use a smoke machine. Will that be safe?

....

Answer: probably, but outdoors

I have had a child by IVF. I was put on the immunosuppressant dexamethasone during the procedure. I want to have another child, but they are refusing me the drug because they say it will suppress my immune system and I might catch COVID-19. What should I do?

....

Answer: talk to your GP.

I am a separated father of a 15-year-old girl. She lives with her mother. Every year for her birthday I drive her into town and we go to a lovely restaurant and I buy her a nice present. Her mother won't let me this year because she's scared I might infect her. What should I do?

....

Answer: have a test. If negative, drive her into town in your car with the windows fully open, blaring loud music, wearing masks.

I am a convenor of the University of the Third Age. Many of us are over 70. The women really want to go to their hairdressers. They are frightened as they might infect their families. What should we do?

....

Answer: it's OK to go to the hairdresser as long as they are following all the guidelines. Spend a short time there and wash your hands when you get home.

Response: You have no idea how happy you have made so many women today!

....

I am a music teacher. I know singing isn't allowed but would it be OK for the children to hum and use body percussion instead of instruments?

....

Answer: I don't know. But if it's outdoors and with good distancing and for a short time this should be fine.

My grandson was recently taken into care, as sadly my daughter couldn't look after him because of depression. The foster mother is refusing to allow my grandson to visit me as she says he might get infected and bring the infection back to her as she is in a vulnerable group. Can you please email me to say children aren't a major source of infection, as I want to show that to the judge who is examining this case.

....

Answer: I sent her an email confirming that children are not a major source of infection and she got back to me to say my email helped 'carry the day'.

And then some questions I couldn't really answer:

If I carry a jam jar to spit in discreetly when I need to, will that lower my risk of catching COVID-19?

....

I've noticed the wearing of flip-flops and sandals is on the increase. Might droplets land on people's feet and then their feet are a source of infection? I think I'll stick to my wellies in Ireland for now.

....

Do you think wearing a snorkel might protect me from infection?

....

I think I caught COVID-19. I took a slime extract from snails in my garden and drank one spoonful. The COVID-19 was gone in three weeks. What do you think?

....

As well as questions, the nasty comments:

When the rest of the Irish people find out you have been lying through your crooked mouth, they might like to do more than send you an email, so good luck with that, professor. Must be worrying that Fauci is being thrown under a bus. Who knows who might be next?

....

What does it feel like to be a brain-dead moron?

....

But far more nice than nasty ones, like these:

Hi Luke. I am a plumber in my 60s. Worked for years in the rough end of the business: drainage, sewage. Worked on oil rigs. Never caught a cold or flu in my life. And listening to you, I know why. I've a great immune system! Much appreciate your wise words. Maith thú!

....

I'm a huge fan. Would you sign one of your old
T-shirts and send it to me?

....

Here is a picture of an echinops [globe thistle].
Isn't it just like COVID-19? Please keep up the good
work!

....

And from the Irish ambassador to Argentina:
Please keep informing us of COVID-19. I'm sending
you information on inhaled ibuprofen, which I
know you have a strong interest in, which has great
potential and is undergoing clinical trials.

....

And from the CEO of Laya Healthcare:
Keep up the tremendous work! We need you.

TUESDAY 7 JULY

It struck me this morning. The way this virus has
changed my working life is substantial. No travel – I
miss that. I also miss giving talks and meeting fellow
scientists to thrash out ideas. I wonder, will that affect
progress? And I hate Zoom drinks. Awkward silences.
Boring stories. And all this media – massive.

Stevie got his results today – a first! So proud of my
lad. How will COVID-19 affect him and his generation?
We all need to look out for them. We went to Rasam

in Glasthule to celebrate. It's always been a favourite of Stevie's. They had screens up everywhere and the waiters wore masks. The front-of-house guy recognised me: 'You're the virus guy!' Lovely atmosphere and lovely food. Good to mark the occasion for my great lad.

WEDNESDAY 8 JULY

Downloaded the new COVID Tracker app today. It's good! Over 700,000 other people also downloaded it. This should surely help. You can check in and find out if you're a close contact with someone who has been infected. It will need compliance. Let's see if it works out.

MONDAY 13 JULY

Didn't do much over the weekend. Reading a super book called *Around the World in 80 Trains*. Some of them sound brilliant – across the US in sleeper trains, or the bullet trains in Japan. It's like forbidden fruit, given that we can't travel. Do people in prison read books on things they can no longer do? It's like we're all in prison these days.

Exciting developments for the Pfizer/BioNTech vaccine. Pfizer are a huge vaccine manufacturer, so it makes sense for them to partner with BioNTech. They have said they will report their big trial by the end of the year.

Every time I am in the Newstalk studios, Cormac, the sound engineer, asks me, 'Any sign of the vaccine?' I can't wait for the day when there is one – not least because it will stop him asking! Pfizer have said they are seeing an excellent antibody response. Meanwhile, Novavax have received $1.6 billion from the US government as part of 'Operation Warpspeed', by which many companies are being supported in their efforts to get a vaccine as soon as possible: 190 vaccines are currently being tested, and at least 263 treatments for people in hospital. The FDA are mandating a minimum of 50 per cent efficacy for any vaccine, which seems achievable. As ever, it won't be for want of trying, and I have full faith that at least some of these will succeed.

Watched *University Challenge* with the family. Stevie and Sam are getting questions ahead of me. How dare they? Also watched CNN late into the night. It's become a habit: I flick all the way up the channels until I get to it. Wolf Blitzer, Erin Burnett. I'm awestruck at the way Erin asks a question and then listens intently, slowly blinking. What's happening in the USA is atrocious. Cases climbing everywhere, followed by hospitalisations and deaths. Where's Levitt now? What is wrong with Trump and the administration? Why won't they listen to Tony Fauci? He spoke about death threats to him and his family. He gets it much worse than me! He has security to protect him. And all he does is talk about data and science. Some country.

TUESDAY 14 JULY

More optimism about Roche. Waxes and wanes. Realised today that trip after trip has been cancelled in the past month: Rhode Island, Boston, Melbourne, Taipei …

WEDNESDAY 15 JULY

Big meeting in the lab about the data on COVID-19 coming from Belgium and Holland. Our potential treatment, itaconate, protects human lung cells from being injured by COVID-19. Our first real data! It's a good feeling. Need to egg them along. This can be tricky with collaborators, as they can drag their heels and be more into their own projects.

Meeting outside Zambrero's, the coffee shop near our building. Ann and Ger from Tallaght Hospital showed us lots of data from COVID-19 patients. They, like other labs, are seeing a big increase in various inflammatory markers. We are keen to measure our favourite marker, IL-18.

Finished the day speaking at a Zoom conference on the consequences of COVID-19 for people who are on immunosuppressants for diseases like rheumatoid arthritis. The bottom line is not to worry but to take all the precautions.

What struck me today is how this virus has touched every single part of our lives – birth, childhood, school, sport, music, jobs, hobbies, friendships, weddings and, as I saw the other day, even people trying to get pregnant. I challenged people in the lab today to name a single part of their lives not affected by it and they couldn't.

The Taoiseach (who is now Micheál Martin) said that further easing of restrictions will now not go ahead until 10 August. I wonder why? I guess the trend in the numbers is changing. He emphasised a few things: face coverings must be worn in all shops; wet pubs will stay closed; a maximum of four households can gather but there can be no more than 50 people gathering indoors. It's not exactly easy to follow, but I guess there is some relief. I picked up a sense that there is fear of another lockdown being needed. How will we all cope if that is indeed the case?

An anxious mother emailed me about her son, who has had Kawasaki Syndrome. She is worried about him catching COVID-19. I linked her up with my Cedars-Sinai friend, Moshe, and he got back to her quickly to reassure her. She was amazed that I had managed to link her with a world expert so quickly. Moshe is a sound person.

THURSDAY 16 JULY

COVID-19 patients have a lot of clots in their lungs, I explained to Pat this morning. The disease is increasingly being seen as a coagulopathy. This means that the virus provokes clotting and that can then damage the tissues, cutting off oxygen. We discussed 'happy hypoxia', a feature of COVID-19, which makes it different from other respiratory diseases. Usually people who have low blood oxygen feel faint and pass out – that is not the case with COVID-19, where they feel normal even though blood oxygen is low. Another mystery of COVID-19. Low oxygen normally leads to high carbon dioxide in the blood because the exchange between oxygen and carbon dioxide when you breathe in isn't happening. This is what causes unconsciousness. It doesn't appear to be happening with COVID-19, hence the happy hypoxia. It could be caused by tiny blood clots, which means less blood going through the lungs and so less oxygen being taken up from the blood. This got a big reaction from the listeners, with lots of questions coming in.

FRIDAY 17 JULY

Klaxon horns needed. Got a text at 6.10 a.m. from Manus, chair of Inflazome, scheduling an emergency board call at 12 noon. Oooh!

Then – the call. Roche are now very keen. They came

in with a new offer two days ago. They are prepared to spend huge money on clinical trials on all the big inflammatory diseases, neurodegenerative diseases, respiratory diseases and beyond. I mean, wow. Just imagine if it works in even one of these! I can't quite believe it is happening

Manus asked us to vote on the current deal and we all said 'yes'. He then congratulated me for getting this whole thing started. To see a science project through to this point (a major offer from a big pharma with the real prospect of helping patients) is a massive milestone. Oh my.

Went for lunch in Dunne and Crescenzi with old college mates Jock and Peadar. Was so beside myself, but had to show restraint. Partly because it might still fall through. Told them we were close but not quite there yet. I've been telling them this for years, but they certainly picked up on my excitement. We had two bottles of wine.

Went back to the lab walking on air. It's a strange feeling to be nearly there: I can see the top of Everest but there are still a few yards to go and we could still slip. A mixture of joy and fear. I'll bet some language has a word for it. Tenterhooks? What are tenterhooks, anyway?

On the way home I went for a walk along the seafront at Newtownsmith. High as a kite, I put in my

AirPods and played 'Lily the Pink' by The Scaffold. 'For she invented, medicinal compound.' All the blue, yellow, red and green lights twinkled that bit brighter. I felt like I was in a Van Gogh painting.

SUNDAY 19 JULY

Went over to Sam's friend Gav's house for his 21st. In the garden, all very nice. But then a surprise: his mother had ordered a kiss-o-gram. She was somewhat raunchy, to say the least. It made for a memorable birthday for poor old Gav!

MONDAY 20 JULY

That Monday morning feeling … the comedown. As Tommy Tiernan says, 'the tide comes in, and the tide goes out'. It certainly went out this morning. Must be the contrast. Last week was so hectic, and then my brain ramped down for the weekend. New week, brain has to ramp up again.

Several antibody therapies are in development and Eli Lilly's one looks especially interesting. Their antibody is a cocktail that can mask the spike protein and stop the virus from getting into the lungs. This is a really good plan B if the vaccines prove problematic. Much more expensive to make, though.

TUESDAY 21 JULY

Alex's paper got accepted today!

Alan English, Editor of the *Sunday Independent*, has said I should submit three of my pieces for this year's journalism awards. I'd sent him five and he said they were all great. I need to decide among 'A nation holds its breath', 'Health versus economics', 'The Covid shark is still out there' and 'Face masks are key to halting coronavirus'. I'll have a think about it.

WEDNESDAY 22 JULY

Wrote a piece on vaccines for the *Sunday Independent*, getting the people ready. Imagine if the vaccine comes sooner than we thought. Joy to the world ahead of Christmas.

Acting Chief Medical Officer Ronan Glynn made some worrying noises this evening at NPHET's press conference. They are seeing outbreaks in various places where people gather – building sites, supermarkets, fast-food places. Could this be the start of another spike? Oh, I hope not. Imagine another lockdown. How will people cope with that?

The government has also published a list of 'green countries'. People coming in from them don't have to quarantine for 14 days because the viral count is low in them. They include Malta, Finland and, perhaps surprisingly, Italy. It will be renewed every two weeks. It all seems a bit half-baked to me. Surely numbers will keep changing? And countries like New Zealand just banned travel and have mandatory quarantine from any country. Could we not think about that? Seems too much to contemplate or organise, because we're in the EU, where there is free travel, and because of the border with the North. I can hear the virus laughing. Still, at least it means some travel can happen between countries.

THURSDAY 23 JULY

A fantastic session with Pat today. Eimear said she still can't get over the reaction. We spoke about possible endgames. Vaccination will be key. Incidence of 'long' COVID grows. This is defined as having persistent symptoms after COVID-19 infection. The symptoms include fatigue, problems breathing during exertion, aches and pains and difficulty concentrating. Other viruses can have similar effects in post-viral fatigue syndrome. Some studies are reporting one third of people will have such symptoms, which can go on for months.

And an amazing statistic – at one point 90 per cent of all children in the world were kept out of school. That has to have long-term consequences. Got the feeling that we'll be living with the consequences of COVID-19 for a long time. In Germany they are looking for volunteers to attend a rock concert next month – the pop singer Tim Bendzko (never heard of him) will perform. They will track people in various ways to predict how likely it is that a virus would spread. Can't see it working, but at least it's a gig!

FRIDAY 24 JULY

Today was tough enough. Not clear why. Just a day when it all seems more difficult. A film crew came in to do some filming, but it went on too long, and I had to skip a meeting. That bugged me.

Alex Whelan gave a talk on Zoom. I told the lab that he was my old lecturer and mentor. It was in his lectures that I first heard of IL-1, the cytokine that became the subject of my PhD and is the target for NLRP3, so Alex deserves some credit for Inflazome! He has some interesting, if slightly wacky, ideas. I encouraged him to send his idea to *Nature Reviews Immunology,* one of the world's leading immunology journals, and that I would tell them to look at it. It was so good to see him, if only on screen. He is in his 80s and still burning bright.

A lot of economic news today. The government approved a €7.4 billion support package, which will cover what is now known as the PUP – the Pandemic Unemployment Payment. Hundreds of thousands of people are on this. It's essential to keep people going, and if it wasn't paid it would lead to a breakdown in social order. They also announced over €300 million for schools, to allow them to reopen. At last, a substantial investment in education! Class sizes are likely to drop, which is something for which educators have been pressing for decades. It's essential that schools reopen. Everyone is agreed on that.

SATURDAY 25 JULY

Shopping in Glasthule with Stevie. Queued outside Noel Kavanagh's butchers in the sun. Funny how feeling the sun on your face makes you relax. Good chat to Noel, who as usual slagged me off about The Metabollix. Deep down he's a fan, I know.

MONDAY 27 JULY

Pfizer/BioNTech and Moderna announced today that they have begun recruiting thousands and thousands of people into late-phase clinical trials for their vaccines. Remarkable progress over the past few months. We can hope. They both use RNA technology. This means using RNA as the recipe to make the spike protein from the virus. It's injected into the body wrapped up in a fatty bag, and then goes into cells, where the recipe is read off and the spike protein is made. The immune system then reacts to the spike protein, making lots of antibodies. No vaccine using that technology has ever been approved. Risky, but it could be brilliant. I'm sure they have a strong reason to try RNA as an approach.

Some of the symptoms of COVID are well known: most people are familiar with a loss of smell and taste, suggesting that nerves are being damaged in the nose and tongue. But some symptoms are not as well known, like 'COVID toe', where a toe will become inflamed, like chilblains. Still not at all clear why or how that's happening.

TUESDAY 28 JULY

Kept awake by Inflazome matters last night – a rare occurrence for me. Complexities of the deal with Roche were on my mind. Spoke to Manus early this morning and he assured me all was well. It's getting closer and closer. Tantalisingly closer.

WEDNESDAY 29 JULY

Had a magical afternoon. Stevie came in by boat again and picked me up. We had a scoot around on the Liffey and then tied off beside a coffee place near the Beckett Bridge. I told him the latest on the Roche deal. We dropped by my lab and then headed home by sea. It was a lovely evening with a fair few boats out. Slight chop, but we like that. Felt like the luckiest guy alive … (waiting for the trouble to arrive, as John Cooper Clarke would have it).

THURSDAY 30 JULY

Interesting evening! Took part in a cocktail-tasting session with Oisin Davis. All the stuff was sent over: glasses, shaker, mixers and of course the various cocktail ingredients – vodka, champagne … I was the 'celebrity' guest, FFS, but the others were nurses from various hospitals. Connected by Zoom and then Oisin began showing us how to mix. It was great craic. I gave some to Marg, as there were two helpings in each. It was a lovely session and the nurses enjoyed it too! Can't beat a little bit of what you fancy. Oisin said it was a way of keeping his business going, as he's involved in the promotion of Irish spirits and mixers. There was even an elderflower cordial using elderflowers from County Wicklow.

I stayed up for a bit more and went over to Brian's. Enda and Colm, two of the Metabollix medics, were there. Enda told us about his time in St James's at the height of the COVID-19 attack, and how they had managed pretty well, although he got infected himself. He said he felt bad about this, as it was him who was training the nurses how to use PPE! Just shows you how easy it is for our healthcare workers to get infected.

He seemed in good form. Medical rivalry couldn't stop Brian from winding Enda up: 'Sure the ICU is empty now!' Brian had actually gone back into medicine for a few months to help with A & E, and overall, he said it was actually pretty quiet, but the risk of being overwhelmed was always there. The hospitals really have performed very well overall. Strange times indeed.

FRIDAY 31 JULY

Sam told me about some of the haters coming at me on social media. It kind of disturbs me to know that he sees that kind of thing. I needn't have worried. 'Dad,' he said, 'this means you've made it!'

So, July ends. Big Inflazome action. Quiet in hospitals – here's hoping it's not another calm before the storm. Surely in the next couple of months this will start to resolve, because of all the hard work going on in trying to control spread? Pfizer and Moderna recruiting hugely. Onwards.

AUGUST 2020

SATURDAY 1 AUGUST

Today was wild enough. Roche rang early (on a Saturday! Still, what else would they be doing?) to say they had all the safety data, but that they needed to do one last check.

SUNDAY 2 AUGUST

Good chat with Manus on the phone today. Inflazome doesn't do weekends. Roche have said they are happy. Might the deal happen?

We drove to Roche's Point in Cork. Been going there for years with the kids, to Desiree's house, which is near the lighthouse. It's a great spot and the kids have always loved it. Towed the boat down without too much trouble. Marg is such a good driver. Good job, seeing that I can't drive …

Our friends Duncan and Vivienne had a BBQ for us on arrival. It was a lovely summer's evening. Roche's Point is idyllic enough, it has to be said. Right on the sea, and you can watch the big ships come in and go out. No cruise ships this year, of course – another COVID casualty. I tried hard not to think about COVID-19 or Inflazome. Some gin and tonics and a bottle of wine helped.

Couldn't sleep, though. Did some binge-watching of various news channels. Europe is on the move. Everyone seems to be going on holidays. Greek islands, Spanish beaches and French vineyards are full of tourists. This can't end well – I mean, the virus is still out there, for crying out loud. Don't they know? This could well turn into another nightmare, and I can't help but think it will.

MONDAY 3 AUGUST

And so, the holidays begin. Sore head. Chatting with people on the street. Hearing 'You're famous now!' unnerves me slightly. We dropped a heavy anchor as a mooring for the boat, then towed it to Aghada Pier to launch it. Stevie and I drove it back over to the Point and moored it. Job well done and there it is now, bobbing away with the other boats. Had a guitar session with Sam and then played poker. I won a big hand to everyone's surprise, including me. Well, of counters, not money – but still!

The publicans are asking for a special compensation package if pubs can't reopen on 10 August. This is another industry that has been hugely hit by COVID-19 and they are being especially restricted, so this is surely justified.

TUESDAY 4 AUGUST

Into Cork city by boat for Newstalk. Took about an hour, and what a lovely journey. Some contrast to getting the Dart into Pearse Street and walking up to the Newstalk studios through Dublin city centre! Misty morning fog. Past Cobh, around Spike Island and Haulbowline. Past Monkstown then across Lough Mahon to the banks of the Lee. Right into the centre of Cork like an arrow. Great being on the water.

With Jonathan in Republic of Work, where Newstalk has its studio. He said to watch out for the haters, that it could get inside my head. Good advice.

Lots to talk about today. Another vaccine is on the horizon, this time made by Sanofi and my old friends in GlaxoSmithKline. The big 'recovery' trial in the UK was also a hot topic. This is a therapy trial for severely ill patients. It has already come up with dexamethasone.

But hydroxychloroquine had no effect. This trial is more believable, as it is double-blind placebo-controlled, which means that neither the patient nor the person running the trial knows who is getting the active drug or the placebo. It's the correct way to do a trial.

On our way back we had some porridge in a quayside café, having missed breakfast. Delicious! We dropped into Cobh on the way back and had a drink in the Quayside Bar. Two pints in the sun. Then zipped back over to the Point.

Watched the Channel 4 documentary on COVID-19 – I was interviewed for it a month or so ago. 'There you are on Channel 4!' said the neighbours. It was dramatic stuff – done as a timeline from the start. Very informative and disturbing. Depicted really well how it took off in the UK and how the death rate just climbed and climbed.

Disappointingly, but understandably, saw on the news that the government won't be allowing any further easing on 10 August. Pubs will stay closed. They're removing countries from the Green List. The limit of 50 people indoors will remain. So much for the hope of bigger weddings that many were aiming for.

WEDNESDAY 5 AUGUST
Still waiting for word from Inflazome. Matt Cooper, the

CEO, left a message to say that the head of research had signed off on the deal. Can it be true?!

Good old holiday singsong around Dave's campfire. Dave is a great pal in Roche's Point, who is, as he says himself, 'pathologically sociable'. Stevie and Sam played 'Ooh La La' by The Faces together. My two lads, now grown men! Memories of them as little boys on holidays here. And there they are now.

THURSDAY 6 AUGUST

Into Cork again by boat with Dave and Duncan. Very misty and atmospheric on the water. Couldn't go at top speed because of the fog. Appropriately enough, in the afternoon did a conference call with the UK Maritime Insurance Association. I did an overview of COVID-19. They spoke about how the cruise industry is devastated. All ships are waiting in various parts of the world. Loads of people laid off. Shipping industry in big trouble, which in turn threatens the maritime insurers.

Manus texted a one-liner: *Deal nearly signed.* Good this is all happening with me down here with plenty of distractions.

FRIDAY 7 AUGUST

02.39 (note the time): Manus texted to say the signing had begun! Not a hope of sleep. Two last signatures were needed and then €380 million will be paid upfront in cash. This will be one of the biggest biotech deals in Europe this year – and for a little Irish company!

Gill tweeted today about my new book coming out in the autumn – first sight of it.

The Taoiseach said today that Kildare, Laois and Offaly are to go into more stringent lockdown because of case numbers there. This may actually be a sensible way forward – localised lockdowns. Let's see how it goes. People are giving out, but it makes sense.

18.35: Manus and Matt texted: *DEAL SIGNED!* Then they called. We were all almost in tears. Congrats all round. I just can't quite believe it. I was actually in the car when the news came through and I pounded the roof, just like The Dude did in *The Big Lebowski*. And then I put on 'Lily the Pink' – seems tremendously apt. Have we invented a highly efficacious medicinal compound? We'll find out!

My thoughts immediately turned to all my mentors: Charles Dinarello, Jerry Saklatvala and also Jürg Tschopp, the Swiss scientist who actually discovered NLRP3, the target for our drug.

MONDAY 10 AUGUST

Last few days were a bit of a blur, to be honest. We drove

to Wineport Lodge near Athlone, to my old college friend Jane's hotel, to help launch a boat built over the winter by friends Tony, Garrett and Michael.

My old friend Cormac Kilty, who is a stalwart of the Irish biotech sector, said, don't believe it until the cash lands in your bank account. We're sworn to secrecy, as the deal now has to go through a process involving the monopolies commission in the US. This is a huge punishment of course, as I am BURSTING to tell people in the lab and former lab members who were critical to it all – Rebecca, who played a blinder on our key publication, and Seth, who brought NLRP3 to my lab and did some essential experiments.

I just have to banish it from my mind. It's hard though, as this is about my life's work. And the possible benefit for patients! It seems crazy, and yet that's what the science tells us.

Meanwhile, back at Roche's Point, Stevie's friend Ben turned up. He is the image of Jimmy Crowley, which is handy in Cork.

If you're caught without a mask in shops, hairdressers and the like, you can be fined up to €2,500 or face six months in jail. Some turnaround! This should work and is exactly what I want. Masks protect us, simple as that, and anyone who says otherwise is wrong.

TUESDAY 11 AUGUST

I've just about stopped thinking about Inflazome. Now slipping into holiday mode. Days coalesce. Gave a Zoom talk at the Gasyard Wall Féile, a festival founded by a dear old friend, Stephen Connelly, who died tragically young of cancer. Usual range of COVID-19-related questions.

Can't stay away from the COVID-19 news, though. Putin announced today that the Russian vaccine, Sputnik V, is now available for the general public. He said he'd given it to his daughter. This vaccine looks pretty traditional. It has adenoviral vectors to deliver the gene for the spike protein, so it's a bit like others – AstraZeneca and Johnson & Johnson, whereas the Pfizer/BioNTech and Moderna vaccines are quite different, being built from RNA. All of them make the spike protein, one way or another. Sputnik V might well work, but yet again an unusual approach, as normally it would go through a Phase 3 trial first. We'll see. So now there are two vaccines in widespread use, Sputnik and Sinovac. Those Russians and Chinese don't mess around.

WEDNESDAY 12 AUGUST

Well now, today was a special day. About a month ago I was asked to film with John Creedon in Ballinspittle for his RTÉ show on place names. So this morning we got the boat over to the Old Head of Kinsale. Marg, Stevie, Vivienne, Duncan and Dave. It was one of those days where the sea was like glass, sun beaming down. A rarity in Irish waters. So gorgeous, like a blue-and-white dream.

Tied off at a lovely little harbour. Rita from the production team picked us up and we drove over to a house in Ballinspittle for lunch. Ardal O'Hanlon was there and had been filming at the statue, famous once upon a time for moving. The nice local man who owned the house told us about how at the peak of the moving-statue craze there would be loads of coaches parked around Ballinspittle.

After lunch we set to work. My job was to talk about a so-called chalybeate well. These are wells whose waters have healing properties because they are rich in various minerals. The name Ballinspittle actually derives from *Baile an Ospideal* (town of the hospital) because it had at least two healing wells. Tunbridge Wells is a famous chalybeate well in England.

A local historian had found the chalybeate well on the grounds of an old manor house, where in the 1800s the gentry would take the waters. It had gone into disrepair but had been renovated. We walked deep into a lovely forest and stood around the well as a drone filmed us and I explained away. The waters are rich in the health-promoting minerals, which are indeed good for your skin, and the water itself would be excellent for digestion, a bit like Andrews Liver Salts. In times gone by, a major benefit was just hygiene. The analysis did, however, show a bit of *E. coli*, which is found in sewage. The well is far from any sewage system, so this was most likely from animals getting into the water and leaving droppings. A small amount of *E. coli* is good for you (depending on the strain!), bringing benefits to your immune system. John asked me would I drink it and I looked at the data sheet and said, 'Not today!' They gave me a lovely old glass bottle to take a sample of the water from the well, and even put a camera underwater to film! I haven't had so much fun in a long time.

And I have to say there was something magical about the forest. Deep green. Dappled sun. And the whole thing resonated with Inflazome. The ancients needed treatment for ailments as well. Who knows, maybe that water has an NLRP3 inhibitor in it too?

Once filming was over, we went back to the harbour. Dave and Duncan were sitting outside in the evening sun having pints. I joined them and we had another. Someone came over to our table, apologising. She thanked me for keeping her sane during lockdown. We left the pub and headed back to the boat, scooting back over to Roche's Point in no time.

As Lou Reed famously sang, 'such a perfect day'. They come all too rarely, and even more so in these times. Got to make the most of it.

THURSDAY 13 AUGUST

Somehow, we're all hoping COVID-19 might be almost gone by winter. But like other respiratory viruses, it likes dry air. It spreads most of all indoors, which is another reason why winter is a worry, as we all move inside. Our upper airways dry out and the virus can get a foothold. We'll be under pressure if there is still a lot of the virus around. Makes the public health measures and testing-trace-isolate all the more important.

A gang of us headed over to Bunnyconnellan's restaurant in the boats. A little armada. Anchored off and got in by dinghy. Love doing that kind of thing. Great to be eating out! Headed back for two episodes of *The Wire* with Stevie. It's one of those things – you hear a huge number of good things about a film or TV show, watch it and think, *it's not that good.* I couldn't understand some of the dialogue – a sure sign of the descent into middle age! Or not – Stevie couldn't either.

The government announced that they are moving from phases of reopening to a colour-coded system. Yet another system. Chopping and changing.

FRIDAY 14 AUGUST

Had a good think in bed this morning. This is why holidays are so important. Your brain can go to a different place. It takes a few days to clear out all the thoughts that you carry from your pre-holiday life. You can order them, maybe see things differently.

There are four things I want to keep doing workwise: 1. The lab – all the science, data, contribution to COVID-19, but not to forget the other diseases we are interested in; 2. Students/postdocs – still love helping them. I'm an academic after all; 3. New medicines – keep working on that: Inflazome, Sitryx, other companies; 4. Communication in all its forms: lectures, talks, books. Really enjoy that.

Over to Dave's place in the evening. Traditional announcing of the winner of the annual Roche's Point golf open. Dave gave his usual rendition of 'Seaside Golf' by John Betjeman: 'How long it flew, how straight it flew, it cleared the rutty track.' Even I know it off by heart. Simple things are more important than ever. Then we had a good old singsong and Caitlin, Dave's daughter, sang a wonderful version of 'Sunny Afternoon'.

It all reminded me of when I was a child and the grown-ups would come back from the pub and have a singsong. I was allowed to stay up for it from when I was about 12 and I'd watch them doing their party pieces. Didn't know they were all pissed, mind! My dad always sang 'My brother Sylvest', while my mother would sing Judy Garland numbers. They performed with such passion. My Uncle Tommy used to sing 'The Sheik of Araby', with my dad saying 'With no pants on' after every line:

Tommy: 'I'm the Sheik of Araby'
My dad: 'With no pants on'
Tommy: 'Your love belongs to me'
My dad: 'With no pants on'
Tommy: 'At night when you're asleep'
My dad: 'With no pants on'
Tommy: 'Into your tent I'll creep'
My dad: 'With no pants on'

I remember that so vividly, partly because everyone just cracked up, and also because I knew that it was a bit risqué. Yet again, funny how when you're on holidays your mind can open up.

SUNDAY 16 AUGUST
To round off the holiday we watched three episodes of *Breaking Bad*. Just superb! At least I'm not like Heisenberg, who lost out on millions from the company he founded!

MONDAY 17 AUGUST

There is often a baby boom after a pandemic. Will this happen post-COVID-19? Wouldn't that in its own way be something marvellous?

Got a lovely card today from Joe Duffy. Nice of him to write.

I notice that several universities have moved their courses online. Doesn't bode well. We'd better get ready for all teaching being online.

TUESDAY 18 AUGUST

Signed share certificates for Inflazome in my capacity as director. Still feels like a bit of a dream. Can't talk to anyone about it, either, although I hinted to the lab that something was afoot. I would just love to tell them all.

Gave another talk as part of the 'Provost's Salon' series. Lots of people joined. Most notably Lord David Puttnam, who asked some interesting questions. (I didn't tell him Stevie was going to Caius just like Harold Abrahams in his movie, *Chariots of Fire*.) These salons seem to go very well, and people are very appreciative of my sharing my knowledge. Sure, what else would I be doing?

WEDNESDAY 19 AUGUST

One thing that perturbed me on holidays was the need

to recruit new people. I knew this would be a key task on my return. So this is a new mission. Interviewed a possible new PhD student – Shane O'Carroll. He did his first degree in UCC and had just completed our MSc in immunology. He was good! Will check out his references.

Also had a call with our Dutch collaborators. They will now test if itaconate can prevent SARS-CoV2 from replicating. We know it can save the infected lung cells from dying, but it might also kill the virus.

Meanwhile, things are definitely going pear-shaped again with COVID-19. And we were looking so good with the numbers. But they are now going up again, and not just in Ireland. All over Europe. We had almost got rid of the damn virus. But the number of cases and hospital admissions don't lie. Outdoor events now limited to 15 people. Indoor to six people. Everyone to work from home again. Oh Lord, here we go again.

THURSDAY 20 AUGUST

Eli Lilly's antibody cocktail trials are showing that it is indeed effective for hospital patients and might even protect from infection. A trial is now running in care homes in the US to see if that's the case. Antibodies might well have an important role to play in the treatment of COVID-19, and this trial gives more hope.

There was also more evidence that reinfection might be possible. It struck me: will we ever be free of it? I didn't want to say that on Newstalk.

FRIDAY 21 AUGUST

Stevie came in for lunch. We went to the Lombard pub and talked about Inflazome. Feels like we're in limbo.

Laois and Offaly have been released from strict lockdown. This is good, but I can't help but wonder if the whole country will be put into lockdown soon, the way the numbers are looking.

SUNDAY 23 AUGUST

Drinks in Brian's garden last night. Colm came too and played recordings of some of the songs he's been learning. Wouldn't it be good to get a new setlist for The Metabollix, should we ever get to play again?

Took it easy today. Went to see Desiree. She was able to come outside onto the patio but said it was too cold. So it wasn't a long visit, but it was lovely seeing her. Walking home via the seafront in Dún Laoghaire and yet again experienced that weird feeling of people calling my name. One woman asked for a selfie for her mother! Delighted to oblige.

MONDAY 24 AUGUST

Survey of teens in the US shows that 71 per cent are reporting stress and anxiety – and why wouldn't they? Sweden again saying if the death toll in other countries catches up, they will have got it right, as they won't have wrecked people's lives with lockdown. It's not a competition, for God's sake.

TUESDAY 25 AUGUST

Had a wobble on waking. Happens sometimes. My life has changed in oh so many ways. Big one is no travel. Definitely missing that. Then the demands of the media. Could always say no. What sustains me is Inflazome deal and also the new book coming out. So I got out of bed and into the shower and the glums lifted a bit.

Another worrying piece of news today on COVID-19. Definite evidence of reinfection. They knew this because it was a different strain to the first one. Yet again, good Lord.

I bet it will be like other biological traits. Some will never get reinfected, some will and will get ill, and others might get reinfected but will have a mild disease. We can live with that. Science isn't black or white. And in biology, things are often grey. The normal distribution. And this will be no different. The trick will be to explain that to people.

WEDNESDAY 26 AUGUST

Another recruitment interview. I feel like Yul Brynner, recruiting the Magnificent Seven. The new team is shaping up very nicely.

More worrying news – 22 people have tested positive in a meat-processing plant in Tipperary. Yet again, the unfairness of it all. There is now a justifiable spotlight on working conditions in meat factories. One minor but important benefit from this might be improvements for people working in those places and living in crappy accommodation.

THURSDAY 27 AUGUST

Some day! On the Dart at 7.45 a.m. Eerily quiet. In the Newstalk studios we discussed how llamas and camels are a rich source of antibodies for COVID-19. They have unusual single-chain antibodies that are easier to make. Yet another option in the biggest biomedical fight in years.

Prime Time with Miriam O'Callaghan. We were chatting away and having a bit of a laugh and then she said 'Right, let's get serious' and we were suddenly live. We discussed schools reopening. This is the next concern. Will they be a source of spread? How will the teachers fare? I can't stress enough how the risk of opening schools is much less than the risk of the harm it does to pupils not to attend, especially those from disadvantaged backgrounds. Given the consternation that home-schooling caused back in April, they *have* to reopen. There will be infections, but they can be controlled. I tried to be reassuring.

Some more interesting emails and messages: a 13-year-old emailed to say he and his friends were putting together a book of interviews with COVID experts to raise money to buy dogs for special-needs children. He sent me some questions including: *What did you miss during lockdown?* and *I saw you on TV before and after your haircut. Were you relieved to get it?*

A mother emailed me to say that she and her daughter were in a pipe band. Her daughter had caught piper's lung, which is caused by bacteria. I've never heard of that, but she wanted to know when I thought it might be possible for them to go back to rehearsals as they enjoy it so much.

A mental health nurse emailed about her son who had gone on holiday to Greece with eight friends. What advice would I give her now that her son is back? *No judgement please – my own mother gave me enough of that. Us psych nurses survive on black humour!*

And of course the usual haters: *You are constantly spreading misinformation and lies. God's vengeance on you will be merciless.*

And another nice one! Frank Dillon is the head of the Men's Shed Association and asked if I'd give a talk. I replied that I would be delighted to. It's a superb organisation. *When can we go back to our sheds?* he wondered.

And finally, one from this morning: *My wife and daughter were in a coffee shop and saw you come in. You weren't wearing your mask. Is it OK not to wear masks now?*

This was a coffee shop near Newstalk at 8.30 a.m. I replied saying I was half-asleep and forgot to put it on.

FRIDAY 28 AUGUST

Good Lord, this morning was stormy. Wind roaring through the trees. Wet leaves stuck to the pavement. The government announced a compensation package of €16 million for pubs, bars and nightclubs to help them reopen. I can't see that happening for a long time.

SATURDAY 29 AUGUST

Me and Sam went to the cinema in Dún Laoghaire. We used to go nearly every week, usually on a Sunday. One of our big things to do together, me and my lad. We used to mark the films we saw out of ten. Tonight, it was *Tenet*. We got all the goodies to make the most of it. Popcorn. Coke. Very few were there. Just shows you. But it was GREAT. Just being in a cinema. We were giddy with excitement. Movie was hard to follow but we discussed it for two hours after, and we almost figured it out. Checked Wikipedia to be sure – still a head-wrecker though. Felt somehow appropriate, with time moving backwards and forwards simultaneously. Exactly how I feel.

SUNDAY 30 AUGUST

A piece I wrote for the *Sunday Independent* generated some interest. Risk of reinfection. People had been asking me how the vaccine will differ from the natural infection and what the risk of reinfection is, so I wrote about that. Vaccines are not the whole virus, and they have an immune booster. This is unlike the live virus: it doesn't manipulate the immune response (like all viruses do). Although reinfection after the virus of vaccination is a worry, it's not a huge one.

MONDAY 31 AUGUST

A new person started in the lab! Christian, a new PhD student. Spent time with him, explaining how my door is always open. And also went through the ideas I have for his project. Big day for him. I was reminded how almost exactly 25 years ago I started my own PhD. All those years ago and look at me now. We all journey

KEEP CALM AND TRUST THE SCIENCE

through this strange thing called life. I wonder what Christian's life will be like in 25 years' time? Now there's a thought. We need that time machine thing in *Tenet* to find out.

And so summer ends. August is over. The melancholy of September beckons. It must be said, August was something else. Inflazome. And COVID-19 coming back again: stay away, we don't want you.

SEPTEMBER 2020

TUESDAY 1 SEPTEMBER

Another month begins. It's all beginning to feel more and more like *Groundhog Day*, COVID-19 version: Wake up. Shower. Work. Home. Dinner. TV. Bed. No big nights out to break the monotony. No trips away. No spontaneous gatherings with different friends. No spontaneity at all, so every day is like plain vanilla. How will we keep going like this?

Thank God I'm a scientist. The good thing about science is that 'new shit keeps coming to light' as Jeffrey Lebowski says in *The Big Lebowski*. Why am I thinking so much about movies? Cinema longing, obviously. Here are my top five: *Jaws* (three movies in one: horror/political/buddies); *Withnail and I* ('Ah, Baudelaire!');

Lebowski (obviously); *School of Rock* ('Raise your goblet to rock!'); and *The Commitments* ('The Irish are the blacks of Europe').

In the lab today we said a socially distanced goodbye to Sarah. She was a superb part of my team. Off to Galway to beat cancer via natural killer cells in the immune system. Gave her a *Candemic* T-shirt. Great having a son making merchandise – makes leaving-gifts easy.

WEDNESDAY 2 SEPTEMBER

Another busy media day: the *Six One* and *The Six O'Clock Show*. Schools reopening was top of the agenda. All over the country you can hear the sound of banging and hammering as schools are getting ready to reopen safely.

Had another session with the senior management in MTV. Chris McCarthy said he wants to pay me for my efforts on their behalf. If they pay me personally I'll have to give half of what I get in tax, so I had a brainwave: they could make a donation to my research instead. They are happy to do that, which is very generous of them, as I only did a few sessions with them. It will help pay for one of the PhD students, who I will designate the 'MTV Scholar'. I wonder if they will provide a lab coat with the MTV symbol on it? Had a great session with them. There are several working parents in senior management there, and they had lots of questions about schools reopening, home-schooling and all sorts. These people are mainly in the US, which just shows you, everyone has the same concerns.

THURSDAY 3 SEPTEMBER

Today scored high on the 'recognition in the street' index. Walking up to Newstalk from Trinity, three people said hello. One of them shouted from across the street: 'When's the vaccine coming?' If only I knew. The quickest vaccine in history, which was for mumps, took four years and was approved for use back in 1967.

I must keep reminding people that the Oxford/Astra Zeneca vaccine and the Pfizer/BioNTech and Moderna vaccines all use brand-new technologies that should be very powerful against this virus, so we are definitely hopeful of it taking less than four years. I think that March next year is a reasonable timeframe to start the roll-out. Bit better than four years, anyway.

On Newstalk we discussed where the virus might be from. It could have started in Vietnam or Laos, and was brought into China. Yet again, an unknown that persists. And a new era of testing may be upon us as the company AbbVie announce a $5 antigen test. This measures the spike protein in the virus using an antibody (an antigen being the thing an antibody binds to). Hope the government will act on this.

Fascinating to think that primates like gorillas and orangutans, as well as bottlenose dolphins, might be susceptible because they have an ACE2 (the lock that the viral spike key goes into) just like us. Fungie, beware!

FRIDAY 4 SEPTEMBER

Get this! I'm headed over to Galway today for a scaled-back arts festival. The exotica of a trip all the way to Galway. I'm taking part in a debate tomorrow on you-know-what. Marg drove, and it felt weird enough. Not many cars on the road. But the excitement! We stayed in a hotel! Felt different, of course. A strange atmosphere. The bar and restaurant were somewhat full but with gaps between the sets of people. Hand-sanitisers everywhere. Had to schedule when we wanted a table for breakfast.

We had dinner in the restaurant and met Caitriona Crowley, the convenor of the session, in the bar for a drink. Went to bed feeling bloated – gorged myself on the dinner and then had four pints. Honestly, I was like a dog off a leash.

SUNDAY 6 SEPTEMBER

The debate was good. Dave McCullagh chaired. Paul Moynagh and Catherine Motherway (ICU doctor in

Limerick) spoke. Paul said definitively that the lockdown was the only thing to do.

There was a protest today near the venue with people objecting to lockdown and masks. What don't they see? What do they think should be done to help people? I hate negativity. People who just complain or protest with no positive suggestions. It's OK to be realistic and negative, of course, but there should always be a counter to that.

But then I had my first confrontation with a protestor. I was chatting away to Dave and Paul after the event and one of the organisers came and said there were protestors outside waiting to talk to me. *This could be interesting*, I thought. Left the venue with Marg and two people came up and started shouting at me. One was filming and the other was holding up pictures of what appeared to be children who he claimed had been harmed by vaccines. He was jostling me as I walked. We walked to the car and got in. He began banging on the window, face full of hate, and then stood in front of the car. He was shouting at me and accusing me of not talking in the media about the harm vaccines do. We threatened to call the Guards. He left eventually. Maybe he's in pain because he believes that someone he loves has been harmed by a vaccine. Love turns to hate.

We picked up Paul and drove into the centre of

Galway to get something to eat. I was bothered mainly by the man's contorted face and the spit coming out of his mouth as he yelled. Walking through a crowded street, I had a feeling of trepidation. Maybe he followed us? This was a natural response, of course. A threat is supposed to put us on our guard. Paul said not to worry in the least, as most people are cowards. I soon calmed down, sitting outside a pub and having a pint.

Drove back to Dublin in record time as not much traffic at all. Eat our wake, anti-vaxxers!

MONDAY 7 SEPTEMBER

Just how many people has COVID-19 killed? It's so hard to get an accurate number. And excess deaths are complicated, as people might not be dying from, say, flu, but might be dying of cancer because of a lack of treatment. Worldwide, the number of deaths from COVID-19 is 413,000. But Russians have reported a strong antibody response for Sputnik V.

Big call with our Belgian collaborators today. We went through their data. Itaconate does seem to be protecting the lungs of the infected hamsters. One experiment, but hopeful. The only trouble now is that Laurens, the scientist there who did the work, is moving on to a new job so it might be tricky to get someone else to repeat it. We may need other options. Reminds me of a couple of months back when I was approached by immunologists in the National Institutes of Health in Bethesda, Maryland, who said I was welcome to come over anytime to do experiments there. They have been given loads of funding and they are 'hoovering up talent', to quote. The analogy they used was the US recruiting rocket scientists from Germany in World War Two.

I may well take them up on their offer. The key thing is for us to make progress in our research, so I might send some of the team over. We'll see. It's better if they did experiments for us, as it would take time to train up the people from my lab over there. Sometimes research can be about tactics and interpersonal relationships as much as anything else.

TUESDAY 8 SEPTEMBER

Another hectic day. Call with Inflazome at 9 a.m. All the faces on Zoom. Stilted conversation. Odd feeling, as there is still a remote chance of the whole thing falling through. I shared what I thought was a great analogy: it's a bit like we've won gold in the Olympics and we're now waiting for the drugs test – only in this case we've passed the drug test and now we're waiting for the gold.

Leaving Cert results came out today, with 61,000 students given results based on calculated grades. A

Department of Education official said they had 'split the difference' between grade patterns in previous years and grades estimated by teachers. Overall, grades were higher than any other year on record. All very strange for the students, I should think.

WEDNESDAY 9 SEPTEMBER

Something came to me this morning. Following that attack from the anti-vaxxer, my commitment to promoting vaccines is strengthened. Any effort to decrease vaccine hesitancy is worthwhile. My new passion. See what you did, anti-vaxxer? You've made me even more of a vaccine zealot.

THURSDAY 10 SEPTEMBER

There is evidence that the heart can be badly damaged by COVID-19. One scientist put it brilliantly – 'slashes at heart cells'. Yet again, not the flu. And a study on how a supercomputer at a laboratory in Oak Ridge, Tennessee, was given the task of assessing everything we know about COVID-19 and was asked what the best thing in the immune response to target would be. It examined 2.5 billion data points, and it came up with the number 42. Joke! It actually came up with bradykinin. My PhD supervisor, Graham Lewis, who sadly passed away a number of years ago, did important work on bradykinin in the 1950s. Bradykinin has all kinds of inflammatory effects and can damage blood vessels and cause clotting, so this makes sense. There are drugs that can block it and they will now be deployed. And all because of a supercomputer. Who needs humans, eh?

Had a chat on Zoom today with Jane Ohlmeyer. She is standing for Provost, and I am on her team. Jane is tremendous. Got to know her when she was head of the history department at the same time that I was head of biochemistry. I used to love those meetings of the heads, as it was a feeling of 'we're all in the same boat'. You didn't need to worry about sucking up to superiors or being a leader to the underlings. It really showed what a university is about. All these different disciplines but with one common goal: education, research, enlightenment.

Trinity has flaws, as has every university, but where we win is in collegiality and a shared vision. Jane asked me to canvass a few people. How do you canvass people when you can't meet them? Zoom canvassing. It will certainly be like no other provostship campaign in 418 years. All those students who have come through the front gate. Generation upon generation. Jonathan Swift trod these cobbles. And Oscar Wilde, Bram Stoker, Samuel Beckett. Look at all that knowledge that has accumulated in 400 years, though no knowledge of immunology or viruses back then. In spite of all the knowledge, this virus has brought us to our knees. All that learning will now be used to beat it, of that I have no doubt.

Finished the day with a Zoom seminar at Wake Forest University in North Carolina. A lot of people tuned in, including my old friend Professor Cash McCall. Cash is in his 80s and still doing interesting research on bacterial sepsis. Keep on rockin', Cash!

FRIDAY 11 SEPTEMBER

Rory came to film from Newstalk. He's a regular in our lab now, and I get to say the opening and closing lines for my slot with Pat at last! My work with Newstalk has ramped up to twice a week, which is then put into a podcast. Rory films the promo clips and always does an imaginative job with images. Each one looks so cool. Marah from my lab was on one recently – I was chatting to her about her work. She saw a bit of hate under it. We told her not to worry. Times we live in.

Rory is so polite in the lab. Such a sound (and film) guy. His partner works on *The Vikings* in County Wicklow. He was telling me she is sick of being tested twice a week, which is the policy there. I said maybe the new saliva test that is being developed will help, given how invasive the current test is. He said he'd been dragooned in as an extra a few times because of his red bushy beard. They had, however, dyed it black for authenticity, as Vikings didn't have red hair, apparently.

SATURDAY 12 SEPTEMBER

NPHET announced that only two households should mix, with no more than six people. They said this will be in place for months. For crying out loud.

MONDAY 14 SEPTEMBER

They have found 29 proteins in the SARS-CoV2 virus that have not been seen before. This should help us develop new therapies that might target one of these to kill the virus. This kind of thing illustrates how little we know about this enemy, but how we're learning all the time. How can you dismantle a bomb if you don't know all the parts?

Big call with charity Versus Arthritis today. Worked for them for years, chairing their grants committee. They have announced a 'pain challenge'. There is a huge need to develop better medicines to treat pain that aren't addictive. They have allocated £24 million for this and want the best ideas to come forward for assessment. So many diseases keep burning away and shouldn't be neglected because of the focus on COVID-19.

Another late-in-the-day seminar, this time at

the University of New Mexico. I couldn't resist the opportunity to make a slide with Walter White and his definition of chemistry on it – 'the science of change'. Not sure if they appreciated it or not, but since me, Sam and Stevie are watching a few episodes a week of *Breaking Bad* I felt it was appropriate.

Came home and cooked Mongolian lamb with the leftover lamb from yesterday. Well, I say Mongolian lamb. The recipe is a bog-standard stir-fry and you add some soy sauce, sherry and brown sugar. Easy. We could pretend that we were in Mongolia. I almost mocked up a Mongolian tent in the kitchen from a couple of duvets. Sam could have made yak noises. The thought of travel yet again is tantalising and disturbing in equal measure.

I only tormented myself further by reading more of *Around the World in 80 Trains*. She's reached Japan and is whizzing around on the bullet train. Bah.

TUESDAY 15 SEPTEMBER

Signed around 50 documents for Inflazome today. My signature isn't great, it has to be said. A bit like a big 'L'. I said to a colleague, well, I am like the Queen, who signs with an 'E'.

In a rather disturbing development for the whole country, the entire government has been told to restrict their movements because Minister for Health Stephen

Donnelly felt unwell and may be positive for COVID-19. Now there's irony – the health minister feeling unwell. And the whole government is in quarantine. What next, I ask?

WEDNESDAY 16 SEPTEMBER

Minister tested negative so the government is released. Err … phew.

There are now 30 hours to go for the clock to wind down on whether there will be an objection regarding Roche acquiring Inflazome. Not that I'm counting. Ghosting at midnight tomorrow will mean the deal is done.

THURSDAY 17 SEPTEMBER

Today the lab was supposed to be in Poland. We had booked the flights and accommodation to go on a lab retreat to Torun, the place where Copernicus was born. It was to have been a scientific retreat, but also to celebrate Z's paper being accepted in *Nature Communications*, and also S's paper in *Cell Metabolism*. It's so important to celebrate these things, as it's such a huge effort to make any kind of discovery and get it published. These are two important papers, so our plans were well justified. And of course, because Z is Polish it was wholly appropriate (and cost effective!) to go to Poland.

We had been looking forward to the trip so much, especially Z, who is proud of his homeland. And it would have been great to go to where Copernicus – or as we used to call him in school and titter every time we said it, 'Copper Knickers' – was from. The astronomer in many ways started the scientific revolution by proving that the Earth goes around the sun rather than the other way around, causing consternation, rebuke and denial. Still going on today when you look at people denying COVID-19. The fact that the Earth goes around the sun isn't a political issue. Just like the wearing of masks isn't either. Anyway, our trip is yet another COVID casualty.

It was S's last day in the lab, so we had another farewell. Sad to see him go too – a talented immunologist with a bright future ahead of him. He will be starting his own lab in the University of Graz.

I also signed some last papers for Inflazome. I asked the solicitor who brought the papers to me what would happen, and he said well, nothing. You just don't hear from them. I asked him to confirm to me at midnight that they had heard nothing.

Meanwhile, more messing with travel. Two countries added to the Green List, seven removed. This isn't the way to do things.

FRIDAY 18 SEPTEMBER

As the clock struck midnight I waited for the message. And waited. And it didn't come. Fell asleep.

Then at 9 a.m., while I was scrolling on my phone, I got a text message saying I owed Inflazome €145 and needed to transfer that quickly. Surely the whole deal wouldn't collapse because I forgot to transfer €145? Or maybe I would not get my share? With trembling fingers I managed to do it from my phone. I then emailed to say I'd transferred the money, adding: *Any update?* The reply: *Everything is going according to plan. You'll hear from us this afternoon.* I showered in a daze.

I met Cormac Kilty for lunch in 64 Wine in Glasthule. This was set up a couple of weeks back but was mind-blowingly appropriate. My whole journey into commercialising research started with Cormac. At least 18 years ago an old colleague, Dermot Kelleher, Cormac and myself went for dinner in the Guinea Pig restaurant in Dalkey to discuss setting up a new company around our research. Cormac had years of experience in biotech and was CEO of a company called Biotrin, which, of all things, at that time had a diagnostic for viral infections in dogs. He then became chair of the first company I was involved in, Opsona Therapeutics.

Over lunch I told Cormac the deal was in all likelihood done. First person to hear that. Then I got a text from Diarmuid O'Brien, TCD's chief innovation officer, to congratulate me. How had he heard? Still no official word to me. And then in rapid succession two more texts came in. One from Manus: *DEAL DONE!* And then one from Matt saying that the transaction had been completed and that funds would transfer that afternoon or possibly Monday. Yesss!

En route to Galway for another COVID-19 session, this time as part of what might be called a poor man's Electric Picnic. I'm taking part in a small approximation of MindField, the EP spoken-word stage, called 'The Big Think'. The Metabollix would have played EP again this year. Bah!

Got the Dart to Connolly and then the Luas to Heuston. Played 'Lily the Pink' on repeat. On the Luas, I couldn't contain myself and shed a few tears. They were of joy to have made it but also, I was thinking of all the patients who might benefit from our new medicines. The journey towards that just got faster, with a superb backer in the form of Roche.

I imagined I was in a bar somewhere with all the Inflazome team: Matt, Jeremy, Angus, David, Thomas, Ree, Sarah. And our investors: Manus, Marco, David, Bart. And we're all banging the table to the beat of that song and smashing our glasses together. Except I'm on the Luas.

At Heuston my phone pinged. Jo from Inflazome had sent a video of 'Celebration' by Kool and the Gang, except she's pasted the faces of Matt, Thomas and me over the singers. Got on the train to Galway in a complete daze.

I then began checking my bank account (Manus said to do that) and at exactly 4.12 p.m. the digits in my current account went into overdrive. I took a screen shot and then WhatsApp-ed Marg: *You'd better sit down.*

Manus texted: *You did it! Whatever else is said, you can tell yourself that you did it!* I texted back: *Six diseases will fall to our sword.* Easy to lose the run of yourself but hey, why not? And what if it's true? What then?

Got to the hotel in Galway and went to my room. Strange feeling. Got a message to say that other speakers had cancelled as there is now a travel restriction in place again. Looks like I escaped Dublin on time. We had official letters stating we were on COVID business (public information) but still, some were reluctant to travel. I feel like a pioneer, boldly going and glad to be doing it. I had another moment of emotion. My mind is racing: what next?

SATURDAY 19 SEPTEMBER

Today was good. Woke up and thought *hmmm, the world has a different colour now.* Went to Salthill for the event and met Maria, the organiser. They had rented the driving range in Salthill and put up lots of glamping tents and a bar. They were open each weekend and people could stay in the bubble of their tents, have a few drinks and get some food. In the middle of the field was a big circus tent with the sides up and a stage. It all looked brilliant. Maria is an events organiser, famous for the Galway Oyster Festival. She spoke about how events had been decimated, with lots of people out of work. This event has employed some people. She had to fight with Galway council but won in the end. So impressive. And a perfect example of that never-say-die spirit in people. You'll never beat the Irish.

She brought me over to Will, who was the sound guy for the event. And what a wonderful fellow he is. He is from the UK and has worked in sound all his life, including at Glastonbury. He looks the image of Johnny Rotten. Told me he and all his crew hadn't worked in months. He's the same age as me and we had a few long chats about music. He reminisced about going to punk gigs in the '70s and coming out drenched from being spat at. He told me all about his ideas for 'green' music festivals once this all ends. He put my mic on and then the event started. There were around 100 people there, under the big top, sides all open, sitting at tables, having a beer. It was very convivial. Lots of questions after. Had a couple of pints myself in the evening sun. Drinking in the evening sun – is there anything better?

A brass band came into the field and were brilliant. I listened to them for a while and then headed for the train. All in all a lovely day. It's the perseverance and warmth and consideration of people like Maria and Will that keeps me going. Doing their best for everyone, all of us in this together. Unlike the nasty, negative trolls who are actually harming people, either directly when they attack someone, or indirectly by worrying people or perhaps discouraging them from being vaccinated.

SUNDAY 20 SEPTEMBER

Got back late last night. Cracked open a bottle with the family to celebrate. This morning had a rebound into the glums. We are strange creatures, us humans. We want something so badly. We get it. And then what? I love this exchange in *Waiting for Godot*:

Estragon: What am I to say?

Vladimir: Say, I am happy.

Estragon: I am happy.

Vladimir: So am I.

Estragon: So am I.

Vladimir: We are happy.

Estragon: We are happy. (Silence). What do we do now? Now that we are happy?

Stupidly, concerns come to mind. First-world problem for sure. Would this success demotivate me? Will the money change me and have a negative effect? Will it affect my sons? What about my friendships? On the other hand – wahey! One of my key life missions, to find new medicines, has been hugely enhanced. Get a grip, Luke.

But in the background, I'm worrying away. We're now in Level 3 lockdown, which means all tourist places closed, no indoor gatherings of any kind, everyone in Dublin has to stay in the county, restaurants are takeaway only – all to stay in place until at least 9 October. It's all about lowering the cases in this second spike. Looks like it happened because of people travelling back from Spain, where COVID-19 is widespread.

MONDAY 21 SEPTEMBER

Up at 7.20 a.m., as usual. The press release on the sale of Inflazome was issued this morning. Put Newstalk on and there it was in the business news: 'Irish biotech sells for €380 million upfront plus significant development, regulatory and commercial milestone payments. Founded by Matt Cooper, not of this parish, but also Luke O'Neill, who is of this parish, it's one of the biggest – if not the biggest – sales of an Irish company this year.'

So it's out in the open now. At last I can talk about

it. This was a day full of congratulations. Pat mentioned it at the start of my usual slot, meaning I could point out that the sale means our drug will now go into six diseases, so it's good news for patients too. Eimear the editor was in the studio in person today and said, 'You're rich! Watch out for all the young ones that will be coming after you!' A lot of positivity through the day. The sale will benefit so many sectors in so many ways: my research, Trinity – and most of all, hopefully, the patients. And emails from patients began coming in too:

I just want to say a big thank you. I have ulcerative colitis and I've been on so many medicines and still not symptom-free. To know there are people like yourself who put so much time and energy into developing relief for people like me is just fantastic.

....

Thank you for the hope you bring to so many people diagnosed with life-altering diseases. Wishing you continued success as you move to the clinical trials stage.

....

Well done on your amazing work to place Inflazome's technology on a great path for reaching the clinic and bringing therapeutic benefit to a wide variety of patients in need.

....

What strikes me is that with such a strong following wind we have to have hope that our drugs will work. Roche certainly believe in it. Everyone wants to block NLRP3. So even if our drug doesn't make it for whatever reason, someone else's might, and that would be just fine. That's the way research works. What it all does is give hope to people with diseases, and that is so important too. It's amazing to think that this all began when I had a chat with Matt in the bar at a conference back in 2012. Out of such chats, interesting things can come.

Will Goodbody came in from RTÉ and filmed in the lab. He had been here back in 2015 as science correspondent when Rebecca's first paper on this was published in *Nature Medicine*, so a satisfying bit of symmetry there.

Despite all this excitement, COVID-19 is never far away, and I got another email: *Please don't get distracted working on Parkinson's disease. You have to keep working on COVID-19!* There ain't no pleasin' some people ...

And then a second amazing thing happened – I went out to the Gill warehouse to sign 200 books, just in from the printers. The cover looks brill – same colours as the Sex Pistols album.

Came home tired but happy. Inflazome sale and seeing my book in print. I mean, what the fuck are the odds of that?

TUESDAY 22 SEPTEMBER

And the beat goes on. *Irish Independent* report on the sale of Inflazome. Off to the lab, where graduate student Alessia gave an excellent Zoom talk as part of the Marie Curie training network.

And then more news. It looks like the virus is roaring back. Hans Kluge of the WHO said, 'we have a very serious situation unfolding before us'. He was talking about Europe. Weekly cases are as high as they were back in March. Clearly, far too much travel happened in August.

WEDNESDAY 23 SEPTEMBER

Recorded three lectures for the Trinity junior freshman biology course in the kitchen using the Panopto system. Stevie helped to make sure I was doing the right thing. My fear is that I start into the lecture, talk for 50 minutes and then realise I haven't been recording at all. Luckily, this didn't happen. It's such an absolute bummer that I can't give these in person. I've given the first biology lectures to the freshers for years now. Over 200 students, all eager for their first lectures in university. The human contact is key. How can you inspire students on Zoom?

My job is really to be like a Baptist minister advocating for biology. I try to get them interested, engage them and tell them what it's going to be like in university. I say they can leave the lecture if they want to – lectures are optional because they are grown-ups now. I tell them there are only two grades worth getting: a first (which means they are really smart) or a third (which means they've had a damn good time). It's tongue in cheek, but I think it's important for them to get the idea that they are in a university now and that they can take this subject as far as they want. The point I'm trying to make is that the sky's the limit. Something should happen in the lecture room that will get the student's imagination going. Like live music, I guess.

Anyway, good to get the three lectures in the can. I covered 'The Origin of Life' followed by 'The Chemistry of Life'. What could be better? Got lots of COVID references in, of course, as what else is there?

This strikes me as yet more evidence for how my days have such variety: a discussion with Roche on how our drug might help millions, lectures for 200 of our students in Trinity and then an interview with Spanish TV. I feel a bit like Phil Collins (woe betide!) when he played first at Live Aid in London, and then flew Concorde to play in the US in Philadelphia. 'Funny old world, innit?' is what he said. Poor old Phil is well vilified these days. I'll bet he feels it coming in the air most nights.

THURSDAY 24 SEPTEMBER

There are studies correlating vitamin D deficiency with severe COVID-19. We know it's needed for the immune system, so this makes sense. Needs to be tested further, though.

Recorded my first podcast! Jess, the producer, and I did *The Science of Love*. I really enjoyed it. Me talking for 15 minutes. Nobody talking at me.

Back to the lab. We recorded three little promo clips to plug the book. The book plugging begins (as opposed to butt plugging, which is something else entirely). Then to the podcast studio across the road to record another one with Stefanie Preissner. I am clearly entering the podcast world.

At 6 p.m. Mark Condron from *The Independent* came to take a special photo. He's won lots of awards, so I feel honoured indeed. We projected an image of lungs and COVID-19 on a screen and I stood in front of it. I've realised how tough it is to take a really good photo, and this one was superb. You have to smile, but sometimes you're tired and that must show in your eyes. Marg always says I look like I'm having a stroke in photographs.

FRIDAY 25 SEPTEMBER

Started the day giving a Zoom talk to the European

Society for Clinical Microbiology about our COVID-19 work. Yet another conference that was meant to be in-person.

Z gave his last lab meeting. He's helping Tristram with the itaconate and clotting project and we have initial evidence that it might! This could be really relevant to COVID-19, a disease that causes the blood to clot in your lungs. If itaconate could somehow stop that, it would be brilliant for all infections that feature clotting. Maybe itaconate would be a better anti-coagulant than ones currently being tested? We drew up a plan for future experiments.

We had a socially distanced drink to say goodbye to Z. Another brilliant team member moving on. I'll miss his wonderful sense of humour and in-vivo skills. Great man for the mice, old Z. Dear oh dear, I don't like it when someone like Z leaves, but it's the way of things.

SATURDAY 26 SEPTEMBER

Spent an hour or so today going through the lab 'books'. Who is on what grant, who has left and who is joining. Running a lab is like running a small business. You have to keep the wages coming in. I think I have it sussed for now – enough grant money to cover people for the next two or three years. The job now is to deliver on the research ideas!

Nice sunny day so had a sit in the garden. Noticed a few green, finch-like birds flitting about, twittering. Funny how something that can cut through the worries sometimes, like when a familiar song suddenly comes on the radio. An opening in the mind of some sort, a clearing of the cares we have for a brief moment. Whatever it is, it's great when it happens.

Protestors outside RTÉ last night for Tony Fauci's appearance on *The Late Late Show*. He is being attacked by the far-right and associated people. It's a disgrace. His family are also being threatened. And all he's trying to do is his job – and save lives.

At least my family haven't been threatened, although my sister Helen in England recently got some hate messages along the lines of *You're just as useless as that brother of yours who should shut up*. It unnerved her a bit that they had tracked her down. She's very active on social media, so maybe that's how.

I'm still getting attacks but I'm managing them better. I've cracked Twitter – you only allow people who you follow to respond. And nasty emails are blocked. Such is life now. It's to be expected. If you play rugby, you have to expect to be hit. One thing I didn't see coming was how organised these attacks can be – concerted and mainly politically motivated. I've never been political really, so this is a disturbing aspect.

But I block it. I think Twitter is a hopeless place for debate. How can you debate things in a few sentences? And it allows people to be nasty. It's easy to be nasty remotely. There's a great quote in the *Trainspotting 2* movie: 'Choose Twitter to spew your bile across people you've never met.'

I also recently read that Claire Byrne stopped being active on Twitter when she realised that receiving personal criticism from strangers wasn't particularly useful or healthy. But she is there anonymously. This is exactly the thing to do. Twitter should be mainly about information dissemination. That's its superpower.

As a scientist I'm always debating with other scientists, but in a measured, moderate way. The great European scholar Erasmus said discussion should always be 'temperate', because 'this way the truth, which is often lost amidst too much wrangling, may be more surely perceived'. Cicero said the same when he concluded that the truth is more likely when there is a 'harmonious relationship between interlocutors'. Erasmus knew moderation was best in discussions because 'No man is wise at all times, or is without his blind side.' So true! Mind you, some scientists will still drop poison about you behind your back. They are human after all.

I somehow think quoting Erasmus or Cicero wouldn't do much good for the people who attacked

Tony. Their poster said: 'Rally for Truth. Fake news HQ RTÉ. Tomorrow night, in an act of desperation by RTÉ, "Dr" Anthony Fauci will be on *The Late Late Show* to push the scamdemic and fool the Irish people, as he has Americans for months.' The big question in any democracy is how to tolerate the intolerable while protecting freedom of speech. Not easy, and almost impossible on social media.

SUNDAY 27 SEPTEMBER

Over to Brian's place last night for a good old chat. Could finally tell all about Inflazome. I brought a nice bottle of port. We like port. We necked it in about 30 mins. Followed by a few light ales, as Withnail would say. A bit fragile this morning as a result.

Stevie left this morning for Cambridge. Sob! He drove his own car to the ferry in Dublin Port. Life is so strange at times. I was immediately reminded about how back in September 1985 I had left my dad in Bray to go to London to start my PhD. I went on the ferry from Dún Laoghaire to Holyhead. A well-worn route for millions of Irish. And here's Stevie, 35 years on, heading to England on the ferry to start his PhD. A kind of symmetry, I guess. And those strange feelings yet again of someone leaving who you don't want to leave, but you know has to.

I'll bet there's a single word or phrase for that in Irish – the Irish are great for single words that take several in English. Read a letter from Irish scholar Breandán Ó Cróinín in *The Irish Times* recently, which defined *báisteach leatromach* as meaning the kind of rain 'meant for the guy beside you at a football match but deflected on to you by his golf umbrella'. I wonder is there an Irish word for the kind of sadness that comes upon you when your offspring leaves home. Surely a common event in Ireland over the centuries, so I'm part of a great Irish tradition. Whatever it is, I have it today.

MONDAY 28 SEPTEMBER

COVID-19 is raging in Manaus, Brazil, with cases and deaths soaring. This is what might happen in Ireland if the virus is let spread unmitigated. Pat and I spoke about how interferons could be a great therapy. People who get sick haven't got enough of them. Children make lots. So trials are now running with an inhaled version.

Johnson & Johnson released information on their clinical trial of their one-shot vaccine – 60,000 people taking part. They got $1.5 billion from Operation Warpspeed. This would be a very convenient vaccine, as it can be stored in the fridge. Imagine if we get more than one vaccine, now wouldn't that be wonderful? The Johnson & Johnson one might dominate the market because of convenience.

I then met our latest crop of final-year students who will come to our lab for their project. It used to last 12 weeks but now it's down to four because of COVID-19. At least they'll get some lab experience – so important for their training. We'll do our best for them and stagger them in the lab, one at a time.

Finished the day giving a talk for the Grangegorman Histories Project. Ida Milne gave a fascinating account about the 1918 pandemic from a Dublin perspective. She is part of a WhatsApp group that many working on COVID-19 are in, including me – everyone from medics to scientists to historians and journalists. It's really good and provides updates on many issues. Sometimes though, I've noticed that people on it get a bit nasty towards others. Ida hates that – she recently posted about how academics should always be moderate. She knows her Erasmus and she is dead right. Her perspective on the 1918 pandemic is useful from the point of view of COVID-19.

Worked into the wee small hours recording three more lectures for the junior sophister immunologists. I give them their first three lectures – an introduction to the science of immunology. Obviously, this topic has never been more relevant in all the years I've been teaching it! Immunology is the new ... everything? Maybe virologists would have something to say about that.

One sentence will suffice when it comes to COVID-19 today. One million have died.

TUESDAY 29 SEPTEMBER

Gave a Zoom talk at another conference – the Royal Society for Immunotherapy. Very interesting session on possible new therapies for COVID-19. The steroid dexamethasone is clearly protective for around 20 per cent of people on ventilators. Not bad, but we need to do better.

WEDNESDAY 30 SEPTEMBER

An easier day today. Sometimes it's good when the schedule is almost empty. Time to sit back in my chair and look out the window. And it was rainy today. That sweet September rain.

Another month. By God it was something else. The Inflazome sale. Held my new book in my hand. Stevie off to Cambridge – his life's journey continues. But COVID-19 keeps burning. The storm is descending again. When will we get relief? When? And no obvious vaccine in sight.

OCTOBER 2020

THURSDAY 1 OCTOBER

On with Pat, and we discussed COVAX, a tremendous organisation that aims to provide vaccines to the developing world. So important that there is global access to the vaccines and COVAX are at the forefront of that. Health inequalities that have dogged the world since time immemorial, I would think. Will COVID-19 change all that? The pragmatist in me says 'unlikely', but the optimist says 'maybe, in some ways'. Lively show with Pat, in any case.

FRIDAY 2 OCTOBER

Slight chill in the air this morning. Oh man, winter is coming. And respiratory viruses love nothing better. I shake off my sense of foreboding by thinking about the day ahead.

The *Late Late* was brilliant. I set up a box of dry ice. Added water and the CO_2 gave rise to '80s-style smoke. Just like a Duran Duran video. Ryan then spoke over it with and without a mask. The masks stopped the smoke spreading. It's important to remind people of how effective masks are. We then spoke about the vaccine, and I rammed home the message that vaccines are the greatest contribution to human health above all else.

Just before I went on, I was told that Trump had just been admitted to hospital, so we broke that news on the show. We learned that he had been given the experimental antibody therapy developed by Regeneron, so we talked about that too. I pointed out that Regeneron have a huge plant in Limerick – the former Dell plant. Good to make it vivid for the Irish audience. That antibody therapy could well be the reason why he might survive. I told Ryan he had a 1 in 6 chance of a severe disease, given his age and the fact that he is obese. Looks like they've pumped every drug they can into him, including melatonin. We wished him a good recovery. Never thought I'd be talking about Trump on the *Late Late*, but there you go.

We also chatted about Inflazome, and I was happy to tell people that the trials will begin for the big diseases we are interested in. The money issue came up of course, but I pointed out that Trinity and the exchequer all benefit, and of course the investors.

After my slot I had a nice big cup of coffee in the green room. An almost empty green room, and no booze. Of all the times for me to make it on to *The Late Late Show* – no audience, no green-room craic, and talking again about a killer virus.

SATURDAY 3 OCTOBER

Went shopping in Glasthule and three people spoke to me. I've noticed that as soon as you're not on TV, people forget. Kind of looking forward to that day.

Had a bit of an attack of the imposter syndrome this morning – the feeling that you're not quite good enough. It was first reported in women who had been promoted at work and in successful sportspeople. It's a natural protective measure. Be on the lookout for getting knocked down by a rival. Be on your guard against being full of yourself! I've had it in the past, including as a scientist, say when I've been asked to give a keynote address to thousands of other scientists at a big conference. It passes when I've got on with the activity that is making me feel it. This will pass too. (I hope!)

The feeling reminded me of something else I've come across. I decided to do some research and came across 'do-gooder derogation'. This occurs when someone feels threatened in some way by a 'do-gooder'. The attacker feels they are being criticised by the moral behaviour of the do-gooder. It's like a knee-jerk reaction. So, if I advocate for mask-wearing, a non-mask-wearer might think I'm attacking them, and so they attack back. They see it as defending themselves, or perhaps calling me out as to my motives. I wonder if this evolved as a way to bring down a do-gooder who might have access to resources the non-do-gooder hasn't. If the do-gooder is benefiting the tribe as a whole, it seems counterproductive for them to be attacked, and yet that is what happens.

Do-gooder derogation is made more powerful if evidence can be used to back it up. I've noticed I've been accused of being in the pay of pharma companies, and so there have been charges of conflict of interest. My defence is always that my connections are in the public domain, and that I have no shares in companies developing the vaccines and medicines I mention – and so won't benefit financially. Seems to bug people, though.

Next time I see some hate maybe I should reply with, 'It's OK. I understand. You're a do-gooder derogator.'

That should work a treat. On the other hand, they might just be assholes.

SUNDAY 4 OCTOBER

NPHET have sent a letter to the government, recommending that the whole country go to the highest level of restriction, Level 5, for four weeks. I wonder what the government will do?

MONDAY 5 OCTOBER

The government will not move to Level 5, but instead the whole country will move to Level 3 from tomorrow. Ireland now has the highest number of confirmed cases in a single day since 10 April: 1,205. The second wave is here in earnest.

Leo Varadkar on *Claire Byrne Live* said NPHET hadn't thought through their recommendation properly. Ooh! A spat! Not good for the country. The virus will love that. Varadkar publicly criticising his own advisers … Then to Newstalk, exploring how sniffer dogs are being deployed in airports to detect people who are positive for COVID-19. Man's best friend indeed.

Good three-hour call with Roche – a big overview of all their immunology programmes and how NLRP3 might fit in with them. It gave me a lift. Makes the whole thing so real. There is big optimism there regarding severe diseases. The dream continues. It also got me thinking. I feel really fulfilled when I'm using the skills I've got – in this case, in the area of inflammatory diseases. I guess it's like a footballer being allowed to play. I can lose myself in the scientific discussions sometimes. Flow achieved.

Got a nice poem today written by Con Feighery, immunologist extraordinaire. Con was a stalwart of Irish immunology for years, and when I came back from the UK to set up my lab he was so welcoming. Brings back memories of those early days and the brilliant Irish immunology community: Alex Whelan, Dermot Kelleher, Denis Reen, Cliona O'Farrelly, John Jackson. That first Irish Society for Immunology meeting I went to in St Vincent's – I remember how much fun it was, unlike the stiff meetings in the UK. And now, all those years later, Con sends me a poem:

A badge of serious science
His white coat flung
Across broad shoulders
Long, lean body

In sloppy stance
A lop-sided smile

Dispensing complex information
To an appreciative nation

Or how to perform
A simple task
Wash your hands

Tutorial concluded
Joining the conversation
Time to discard the badge
Not quite Howie in design
Protective gear
Thrown with athletic ease
To the waiting settee

Outside Walter Reed
A medical assembly
In angulated format
All sharply suited
Buttoned-up white atire
The president's commander
Carefully unmasks
Concealed information

Contrasting protection
Contrasting styles

TUESDAY 6 OCTOBER

Read all (well, nearly all) of Alex's PhD thesis last night. Can do it at night because of no interruptions. Took about three hours. He has to submit it soon. It's great! Didn't have to make too many recommendations. So satisfying for him (and me too) to see it completed. He can be very proud of it.

Recorded the next biology lectures. I love doing those lectures, describing the building blocks that make up life itself. Couldn't resist dropping in COVID-19 references. Wonder what challenges this new generation of biologists will face. They'll never forget this year, nor will any of us.

The reviews of *Never Mind* have been good, although in *The Irish Times* Laura Kennedy wondered why I hadn't discussed epistemology more. I almost sent an email saying, *Epistemology is a load of bollocks!* I should have covered it in the introduction, where I discuss the scientific method. It's about how we know what we know, so it's more philosophical. I guess what I mainly wanted to get across was that the most important determinant of scientific truth for scientists is reproducibility. That what is being described can be shown experimentally anywhere, anytime. Epistemologists worry about what scientific truth is. I guess I'm too busy making discoveries to worry about it, but it was fair point.

Then I had an interview with *VIP* magazine! The editor Bianca Luykx came to the house with a photographer. Lots of pics taken, with me holding up one of my favourite albums (*Dark Side of the Moon*). We had a great chat in the garden. Bianca got a bit philosophical when she asked me if I believed in God. I said no. If I'd said yes, do-gooder derogation might have kicked in. Hope the garden looks good in the photos. Divorce is no doubt imminent, which is the curse of being featured in such magazines, apparently …

WEDNESDAY 7 OCTOBER

Culled a few people I'm following on Twitter. Kept some key ones: Thomas Pueyo, Devi Sridhar and EndCoronaVirus.org are all good for COVID-19 stuff. Roger Highfield is a science journalist I love to follow. Every morning at 6 a.m. he goes for a walk in Greenwich Park and sends a photo. When I lived in London, I used to go there a lot. Makes me happy to see the pics. Roger and Devi are great exemplars of how to use Twitter. Keep it professional. And never retweet a compliment – Tony Connelly gave me that tip!

Zoom meeting of the COVID-19 Centre. Great news – we should be able to work directly on the virus itself in the New Year. It will be good to carry out experiments ourselves.

Then a Zoom call with Ann Devitt and lots of teachers. Ann is in the School of Education in Trinity and has a forum with newly trained teachers. Lots of questions came in about COVID-19 in schools, which procedures to follow, distancing, masks, activities like PE and art. I did my best to provide answers, banging home the three Cs and emphasising ventilation and hygiene. I think everyone knows these things, but it helps when they are verbalised.

THURSDAY 8 OCTOBER

There have been definitive cases of reinfection. Someone who had been positive four months ago was then tested again. They detected a slightly different virus, so it had to be reinfection. The question now is how common will that be? I told Pat's listeners it would likely be a range: some will be completely protected, a minority might be reinfected and handle the infection poorly, but the majority might be reinfected but will handle the infection well. It could all be about the dose of virus – a high dose might mean increased chances. Like with most things to do with COVID-19, we don't know.

On 2FM with Jennifer Zamparelli for a book plug and also COVID-19, of course. Odd how the two converge for me.

Tutorial with senior sophister immunologists, three

of them, wearing masks, socially distanced. Good to have some (masked) face time with them. One of them said their granny was so proud to have a granddaughter as an immunologist. Wonderful that grannies are proud of immunologist granddaughters!

Prime Time came in to film. We discussed the numbers going up again. When, oh when, will we have a happy discussion?

FRIDAY 9 OCTOBER

Very, very excited today! Trip to the UK – Cambridge, London and Oxford. Travel is permitted because it's COVID-19 related. I will have to restrict my movements for two weeks when I get back. Dem's the rules. I'll just work from home.

Got a taxi to the airport. Babbled away to the taxi driver with the novelty of it all. Verbal diarrhoea. And to think I used to go to the airport every week almost. I knew I missed it, but I didn't think I would be this excited! Terminal One was almost deserted at 11 a.m. on a Friday morning. It wasn't possible to get the boarding pass onto my iPhone as I would normally have done. Went to check-in desk – flew British Airways (sorry, Aer Lingus!). No queue. One person there. Got my boarding pass. Had to fill in a form giving my address in the UK and Ireland and phone number. Strolled through security. Had my mask on but the woman said 'You're Luke O'Neill. I recognise your eyes!' Had a nice chat. She said only 100 people or so had come through yesterday. One hundred!

Lots of shops were shut and I couldn't get money from any ATM as they were switched off. Boarded flight. I'd say there were 20 people on board. I literally felt as if it was my first flight ever. The excitement. The flight attendant gave me a pre-sealed sandwich and a cup of coffee. Landed in Heathrow 30 minutes ahead of schedule. The flight attendant said it was because there was no traffic in the air so we could take a direct route. Followed a track to the Heathrow Express. Different route from before all this. Train was empty. Got the Underground around to King's Cross. Everyone wearing masks. Silence. Made the train to Cambridge.

The novelty began to wear off at that stage. Felt a bit uneasy. Not many people about. Stations that we went through were all empty. Nobody smiling. A strange silence. I began to wonder, *Should I be here at all?* Got into Cambridge. Ah! My old stomping ground. Spent nearly three years here when I wore younger scientist clothes. Also lived here for six months while on sabbatical in 2016.

Good old Stevie was waiting for me at the station. Great to see him. Had to rush into his car though for a live interview with Kieran Cuddihy on Newstalk … on John Lennon! Seconds to spare. They know I'm a huge fan and it's his birthday today. He would have been 80. Kieran tried to wind me up by saying he only really wrote teeny-bopper music. Good to have a bit of joshing for a change.

Drink with Stevie in one of my old locals, The Flying Pig. Very different from when I was last here. They took our temperature on the way in and then to an outside table. Most tables were full, but not many of them because of social distancing. We ordered our drinks online. Pints of IPA. I thought it would take forever but they were brought to our table. Loosened up a bit with the first, so we had another.

Then to dinner with my scientific collaborators, one of the reasons for my visit. Mike Murphy in good form – I owe him a lot for the recent successes we've had on immunometabolism. Thomas Krieg also there. He is coordinating a big grant we're writing for the Wellcome Trust about metabolic change in heart disease and inflammation. Thomas is indeed a gas man. He's got a sense of humour that matches mine perfectly. We had some laughs. Thomas is a physician in Addenbrooke's Hospital and he told us about treating COVID-19 patients and how the hospital has been coping. He said it wasn't too bad, which was good to hear.

Then back to Mike's for the night. A lovely house outside Cambridge. He produced some port, and we, err ... drank it all.

An excellent meeting. The trip was worth it for this alone as I got a couple of ideas. Shows how face time is so important for scientific discussions. And it's more fun. Perhaps that's the essence of it. If you're having fun, maybe your brain relaxes and ideas emerge. We will go back to all this again one day.

SATURDAY 10 OCTOBER

Stevie drove me back to the station. It's tough enough for him here. He's just started but no social life to speak of.

Train to London. Again, very quiet. Then a taxi to my hotel, The Athenaeum on Piccadilly. Checking in was contorted. Long queue. Form-filling. No room service. Remote control for TV sterilised. Had my second meeting of the trip in the hotel, again on COVID-19. Possible collaboration with a company with some wonderful ideas. Again, the face-to-face was key. We went for dinner in the Hard Rock Café next to the hotel. All the staff wearing masks and 90 minutes table-time maximum. It didn't really work. No buzz. Unease.

Back to the hotel by 9 p.m. What to do? Took a hot bath and had an early night in the huge bed. Very comfortable and a sense of luxury, which is nice. Took it all for granted for so long.

MONDAY 12 OCTOBER

More COVID-19 meetings today with Oxford collaborators: the role of the lymphatics in COVID-19 – an under-appreciated area.

Then to Heathrow. On the bus, something amazing happened. Got a text from Sarah at Gill to say that *Never Mind* was number one in the hardback non-fiction charts! Such a thrill. For the second time this year on public transport I felt a tear well up. They are interesting tears to shed. Joy, but also a sense of achieving something.

I guess the emotion is born of a long struggle, with many failures. I thought about all the hours spent working on the book. And not being fully sure of it. And now there it is. Another strange thing to happen in this year of COVID. Felt maybe like what The Beatles felt in America when 'I Want to Hold Your Hand' got to number one. Not that it's anything like that but still, it's on my mind. More like The Rutles, the spoof band based on The Beatles.

I will cherish this moment. It might never happen again!

WEDNESDAY 14 OCTOBER

Will be locked up for 14 days. This isn't the worst because the government is banning all household visits from tomorrow and the country is going into another lockdown. It's called Level 4. No visitors, six people max at a wedding, all gyms closed. We enter the second big lockdown of the pandemic. For crying out loud.

Got a text this morning to remind me of my quarantine and confirm my location. Impressed by that. Luckily, I can do lots of work from home. Was the trip worth it? Absolutely.

Watched two more episodes of *Breaking Bad* with Sam. Such a good way to relax: drugs, violence, mayhem …

THURSDAY 15 OCTOBER

Johnson & Johnson have paused their clinical trial because of an adverse event. Damn! Creating huge unease. I reckon it will restart, as these things happen in trials. The EU is buying up millions of doses of vaccines from Johnson & Johnson, Sanofi and AstraZeneca. We'll get our divvy from those.

FRIDAY 16 OCTOBER

Took part in another online conference, this time with Tallaght Hospital. Some great talks from doctors there

– one good one on how COVID-19 is a coagulopathy and how they are trying to treat that with anti-coagulants.

A lovely email from Ray Hammond, who told me that he had a summer job in the mid-1960s on Bray seafront, on the deck chairs. He said how he remembered my father: *Now, Mr Kevin O'Neill had an English accent and when I read an interview you gave you mentioned that your dad had come back from England. I wonder, was that your dad? I'm almost certain it was as I remember him well and you are the image of him. He was a kind and gentle man.* Another wee blub! I emailed Ray and indeed he was talking about my dad. He said he'd such good memories – first job, first bit of romance. He also told me that he saw Phil Lynott play in Bray, in Fatima House. A venue I well remember! Ah, memories. Lovely to connect with Ray.

And last week I got an email from John Mooney (my dad's partner in the deck-chair business was John's dad, Pat). John had emigrated to Canada many years ago. He emailed to thank me for all the work I was doing and to say how my dad and Pat would be proud. So nice of him.

SATURDAY 17 OCTOBER

Three name checks in today's *Irish Times*. First, people were asked what they do to cheer up and I said I play

'Monster' by Welsh band The Automatic. It kind of sums up how I feel at the moment when it comes to COVID-19: 'What's that coming over the hill, is it a monster?' Jennifer O'Connell put the piece together and said she loves that song too. It seems to have struck a chord. There was also a quote from my Royal College of Physicians talk and then a bit about *Never Mind* being at number one. *The Irish Times* online also quoted me from an interview with Brendan O'Connor. Good God.

Also got a big box of goodies from the An Post book awards because guess what – *Never Mind* has been nominated for Best Popular Non-Fiction award! I've been nominated twice before and didn't win. I won't hold out much hope. And if I do win, I won't even be able to go up to the podium to thank people.

Cooked smoked ribs for dinner. Something different. Never made them before and we need variety, don't we? They were OK. Sam and Marg were polite, but I know I won't cook that dish again.

SUNDAY 18 OCTOBER

Looking at the week ahead and I've over-committed. I've listed 24 tasks. I find to-do lists great things altogether. I've always made them but as I look back over this year so far, they are all very long. I write them in my diary but also have them on an app on my iPhone. We love to-do lists because we get a dopamine rush when we tick off our tasks. I wonder for me is it about tension-release-tension-repeat? Say I have to do something like give a talk at a conference. I'll wake up that morning and know it's coming. A build-up of tension. Get out and do it. And then release. Our brains are mere biochemical machines, and we just drift along. Sunday rumination. Some year. I wonder how the money and fame (or should I say infamy) will change me?

Went to see Desiree today through the window in the nursing home. Sam came with me. Main topic was, 'You're getting cold out there!' She always thinks of others. Walked home via the People's Park and a guy shouted, 'Love your stuff in the *Indo*!'

Got home and Brian dropped by. I lit a fire in the garden and we had some port. Nice.

MONDAY 19 OCTOBER

Rainy Monday. Interesting chat with Pat on whether vaccines will work as well in older people, whose immune systems aren't as strong. I wonder if it will mean a higher dose of vaccine, as happens with the flu jab? Moderna and Sinovac are reporting a strong antibody response with their vaccines in older people. This is positive news.

TUESDAY 20 OCTOBER

Country now back in full Level 5. Can't go beyond 5 km. Schools still open, though, thankfully. The government were finally forced into it. Why didn't they heed the advice sooner?

Had a call today with Roger Preston in RCSI. (Roger is Ireland's leading scientist in the area of coagulation – good to know that even if we can't meet in person like we used to, we can still set up collaborations.) We have even better evidence that our itaconate molecule can block clotting. This is highly relevant to COVID-19. Itaconate is much lower in patients with severe COVID-19, and maybe that leads to clotting. This could be huge! Many anti-coagulants that can be used carry a risk of bleeding too much. Because itaconate is a natural molecule, made by our bodies, and because it is also anti-inflammatory, it might be ideal. Our company, Sitryx, are interested in making new medicines based on itaconate, so if we can get data on this it would be wonderful.

People keep asking me why I'm so optimistic when it comes to COVID-19. It's because I'm a scientist and deep down, all scientists have to be. Experiments and ideas fail the whole time, so you try again and again because you know you'll get there eventually. Those who give up are the ones who usually realise science isn't for them. That's fair enough. They become tax attorneys.

The night drew in early. Still no news on vaccine trials or indeed no big news yet on therapies. Hang tough now.

WEDNESDAY 21 OCTOBER

The darkness from last night still infecting my mind. But up and at 'em into the shower and the lift begins. Mind you, had to get up at an unearthly hour – 6.30 a.m. – for a Skype interview with Sky News. It's a bit surreal. In the kitchen. Pitch-dark outside. Laptop positioned. Disembodied voice, 'You're on in one minute.' Then the interviewer, but what she says is out of sync with her lips. And I can see my huge head on the TV. Max Headroom.

I'm just astounded by people's kindness. Letters with offers of help in whatever way I want, others encouraging me to keep going with the media to counteract the nastiness that's out there. And people offering their blood for research to make antibodies to help others. One woman wrote: *I have enough antibodies for a whole army.* People just trying to make a difference. To do their bit. Yet again, we Irish show what a decent people we are.

Some more funny letters:

Late March I fell in love, firstly with Tony Holohan but now Luke O'Neill. Thank you so much for all your hard work and sound advice. Now get on your bike and find that vaccine!

....

In conversation I have often heard 'Luke says' and 'Ah sure, he's one in a million'. I must confess I claim you as one of my own, as I am from the garden county.

....

My name is Eoghan. I love you, Luke O'Neill. You're my favourite scientist. I got your *Great Irish Science Book* and I did lots of the experiments. I really want to visit you. I'm seven.

....

I always enjoy seeing you on TV, Luke. I live alone and it can be very lonely. My son lives in Manchester and rings me every night and I'm going over to see him at Christmas if I can. I hope it's the right decision, Luke. Have a lovely Christmas – I will be thinking of you!

....

I am a former Franciscan missionary, now 82. God bless you and the insight you bring. Like when I was on the missions, you too have an important mission and I wish you every strength.

....

And I got a huge card, made by pupils in Carrakennedy National School and signed with good wishes. Good Lord, so kind and a bit overwhelming! I'll write back to them as soon as I can.

The gifts also continue to pile up: bottles of wine, chocolates, endless masks (I'd say 50 or so by now), but also scarves, muffs, CDs and a lovely piece of lace made by a 90-year-old lady. The pattern said 'HOPE'. And at least ten masses to be said for me and two miraculous medals. I am so, so in awe, and so touched by all this. Never in a million years did I imagine this.

THURSDAY 22 OCTOBER

Felt low on waking but perked up in the hot shower as per. Before I went on air with Pat, I tuned in and heard him laying into Minister Donnelly. He always seems to be in the thick of it. Unnerved me a little. We discussed studies showing commercial flights as places where infections spread. On one flight from London to Hanoi, 16 were infected by a single person. Also, an update on interferon as a therapy. A trial has shown a 79 per cent decrease in the need for ventilation. Now that's good news.

Late morning, Brian dropped over to the garden, bringing takeaway coffee with him. Working from home means I miss my fix. Always buy coffee on the way in, usually Tesco coffee, but they've upped the ante with a new variety. I was always struck by how buying coffee on the way into work became a huge trend in cities all over the world. An alien might think that humans have an extension to their hands that appears to be a cup of hot liquid that then disappears in the evening, so it must be retractable!

FRIDAY 23 OCTOBER

Schools have been told to stop using a cleaning product called ViraPro. It contains methanol instead of ethanol. This is a more toxic chemical than ethanol. There's always something.

Interview for the Dublin Book Festival. The promo continues! Sales have slowed a little, but that's because of supply issues. The printer is printing more, yay!

It's quite clear now that Hispanic and Black people suffer much more from COVID-19. Not clear why. Whatever it is, it yet again illustrates how unfair COVID-19 is. It reflects the divides in our society.

Something funny happened in the Spar. I was in the queue and a guy said, 'You're Luke O'Neill!' I said yes. And he said, 'I recognise your red Doc Martens.'

SUNDAY 25 OCTOBER

At around 8 p.m. there was a tentative knock on the

door. And there stood the bold doctors Brian McManus and Colm O'Donnell – a former GP and current neonatologist – with booze. Pissheads! They asked me to join them down the seafront for an al fresco session. It was cold but we braved it and wrapped up well. Colm brought some of his homebrew, which has names like 'Morphine' and 'Amphetamine'. Can't beat medical humour. The sky was clear. Lights twinkling over in Howth. When, oh when, will the pubs reopen …

Passengers on a flight to Dublin infected a further 59 people in the country.

MONDAY 26 OCTOBER

Back into Newstalk after quarantine. The excitement of going into Dublin! The Slovak government are testing the entire army to check the feasibility of universal antigen testing. Johnson & Johnson have restarted their trial, hooray! But AstraZeneca is now on hold in the US. Bah! Pfizer announced that they are making 1.3 billion vaccine doses. Oh, please! This has to be a good sign. They have spent $2 billion so far, have fleets of planes ready to transport the vaccine as well as trucks that can each transport 7.6 million doses. Let's get one of them into Ireland! Very exciting stuff. And holding such promise, but we must wait for the results of the clinical trial.

Gave a talk to Sanofi in France. Big session on immunology. They are developing a vaccine with GlaxoSmithKline, although they said it might have to be tweaked to make it better. Unnerving information. What if the other ones are like that?

We're waiting, waiting. The world is hoping. Of course, the world doesn't know about the challenges. But I am perhaps only too aware of them.

TUESDAY 27 OCTOBER

Had a cracking evening on Bressie's mental health podcast, *Where is my Mind?* Went up to his studio off Camden Street. We had a superb chat. He wanted to cover depression (and my own bout of it), drugs and addiction, but we touched on loads of things. He also asked for The Metabollix to come along and record some songs to go with it, so we did!

Lots of them turned up and it was brilliant fun. Richie on trumpet, Seanie on sax, Brian on drums, Colm on guitar. All masked and socially distanced, so definitely strange. The last time all of us played was in the Dalkey Duck last March at the now-infamous superspreading event, when Ciara was with us and got infected, as did at least ten others. Incredible to think back to that night.

We went for it – recorded 'Don't Stop Thinking About Tomorrow' and 'Johnny B. Goode'. Despite being in three different rooms we managed to get a bit of a buzz going. Bressie filmed a fair bit of it. I'm rusty and have forgotten a lot of the songs on the set list, but hey, if given half a chance … We wondered when the next gig might be. Has to be a huge celebration once we get out of all this, right? I have a dream that there'll be a big concert somewhere, sun shining brightly, and we break into 'I Can See Clearly Now'.

THURSDAY 29 OCTOBER

Pat and I covered the need for humidifiers – keeping the air moist really helps stop the spread of the virus. I told listeners to buy one – they only cost €20. Even a plant that's been watered can be a source of moisture in a room. And jumping in the sea is a top idea! A UK study found that a protein called RBM3 goes up in people who do it. This protein protects against Alzheimer's disease. The shoreline will be full of people plunging in and coming out into their Dryrobes, the number-one fashion item these days. People freezing themselves to feel alive.

Managed to do an in-person tutorial with junior sophister immunology students. Masks, distanced, 45 minutes. To get away from COVID-19 we covered the immunological basis for rheumatoid arthritis, ulcerative colitis and psoriasis. It was really good. Yet again, I'm impressed by students, who are keeping it going in the face of all this.

FRIDAY 30 OCTOBER

Had to spend a fair bit of time working on slides for next week – a big talk to Roche about Inflazome. Thousands might 'turn up' and certainly most of the top brass. It's to tell the company about Roche's latest acquisition. Pressure, as one goal is to explain why they've spent so much on us! It has to be a tight 20 minutes. The head of communications in Roche is in charge and she says keeping to time is paramount. I asked her is it because they are a Swiss company? Worryingly, she thought I was being serious. I want to make my talk really good. It would have all been in person and would have been a big celebration but, as ever, COVID-19 put paid to that. It's a good opportunity for me to gather my thoughts and put a tight set of slides together, as I'm bound to speak on this again in other situations.

SATURDAY 31 OCTOBER

Halloween. Nobody coming to the door. And we didn't put up any Halloween decorations or make a pumpkin head. First time we haven't done that. Might not have done it anyway as the boys are all grown up, but you never know. Saw a good cartoon, a boy saying to his father at Halloween, 'So you're saying to me the one night in the year I want to wear a mask and I can't go out?'

Watched a fair bit of TV this evening. Just flicking and flicking and somehow always ending up on CNN. I know the ads off by heart now. Things looking really bad in the US, with cases rising fast and predictions of many deaths. And of course, the depressing pictures of people just not complying with guidelines.

And so, another month. Ireland is in a bad place. Cases surging. Weeks of being battered now. Winter coming on in earnest.

NOVEMBER 2020

MONDAY 2 NOVEMBER

Great interview with Nadine O'Regan for the *Sunday Business Post* about *Never Mind*. She had read the book in detail and especially liked the chapter on racism, which was a relief. Such a sensitive topic, especially given the situation in the US. I described racism against the Irish to show that we could relate to others. I wonder will it resonate with others?

TUESDAY 3 NOVEMBER

Today I felt fully charged, like the Duracell bunny. I counted it – 11 hours of productive work, not including breaks. Gave a talk to around 200 Google engineers and coders. We discussed mental health issues. Some had read the chapter in *Never Mind* on depression and it had resonated. This is what I wanted to happen with the book.

A key concern was the mental-health effects of working from home. Imagine this virus had come along years ago – home working would not have been possible. People would have had to go to work, increasing the rate of infection. I told them that Zoom has saved lives. So yet again, this pandemic is one for the technological age. I gave them some tips. Take regular breaks. Have a 'commute' before and after the working day, like a walk around the block. This gets your head ready. Something else I read was that working on screens makes things 'transactional', which is not good. We are human beings, after all. It's a poor replacement for physical contact. Many are wondering what this will mean for people, and also for how companies operate. Will it mean a decrease in productivity? We had a lively discussion with lots of questions, so it's very much in people's minds. A lot asking what will happen when the pandemic ends. I said it would probably mean blended working – sometimes from home, sometimes in the office.

Got a phone call just before 6 p.m. Alva from SFI rang to say I'd won the Science Communicator of the Year Award! This made me feel really good. It's great to get feedback that what I do is seen as worthwhile. And the dreaded imposter syndrome is kept at bay. The awards ceremony will all be virtual with no possibility for an in-person celebration. Another example of what COVID-19 has done. No live awards and no need to get up and make a speech. Strange.

WEDNESDAY 4 NOVEMBER

Loved doing the podcast about laughter with Doireann Garrihy – *The Laughs of Your Life*. She asked for my favourite joke. I told her the one about three nuns who tragically died in a car crash. They arrive at the pearly gates and St Peter says, 'Welcome, sisters! You have all lead a blameless life of serving others. But I need to ask you one question before I let you into heaven.' He turns to the first nun and says, 'Who was the first man?' The nun says, 'Oh that's easy. It was Adam!' 'Correct! Please enter,' says Peter. He turns to the second nun and says, 'Who was the first woman?' And she says, 'Oh that's easy. It was Eve!' 'Correct. Please enter.' He then turns to the third nun and says, 'Now, because you are the Mother Superior, I have to make your question a little more difficult. When Eve met Adam for the first time, what did she say to him?' 'Oh,' says the nun. 'That's a hard one.' And St Peter says, 'Correct answer! Welcome!' Doireann laughed and laughed. Can't beat a smutty joke.

THURSDAY 5 NOVEMBER

Did long COVID with Pat, as a report has concluded that it affects one in five. Serious. And a Leipzig study revealed that people shouldn't hang around food vendors. No shit, Sherlock.

Interview in the Science Gallery about *Never Mind*. Eleven years ago, we ran an exhibition there called 'Infectious'. We took saliva from people who attended the exhibition. My lab tested them for a variant in a gene we had found for a protein called MAL, which Adrian Hill's lab had shown might decrease the risk of catching malaria. Little did I think back then that Adrian would be part of the team that would develop the Oxford/AstraZeneca vaccine programme. Imagine how he must feel – making a vaccine that will have a huge impact on millions of people. That's been his big goal for several infectious diseases. He'd plugged away for years on malaria and didn't make much progress, but all that effort wasn't wasted, as he and his team, led by Sarah Gilbert, have done fantastic work. We'll have to put up a statue to Adrian in the Science Gallery.

Today though, the Science Gallery was a bit sad. We did the interview in the PACCAR theatre. I've so many memories of events there, packed full of people. It will come back. The interviewers were really good.

And then in the afternoon – the big Roche talk! No nerves of any kind. It would have been a different story if I'd been there in person. Gave my 20 minutes. Very satisfying to tell the story of Inflazome. I went back to 1986, when I went to a talk by Charles Dinarello (one of my heroes) on IL-1 in London. I began working on IL-1 then. It was clearly involved in so many diseases, so the hope was always there that blocking it would work somewhere. I spoke about a conference I went to in 2001 when Jürg Tschopp first described NLRP3, the key driver of IL-1. And then how we found a drug made by Pfizer (which was pointed out to me by a colleague in Trinity – Tim Mantle) that was able to block IL-1. We showed it worked by blocking NLRP3, and how our paper on that was called a 'game-changer' by Novartis. And then how my co-founder of Inflazome, Matt Cooper, drove it all hard to find compounds better than the Pfizer one, and which were then bought by Roche. I reminded Roche that way back in 1984, they were the first to clone the gene for IL-1alpha, so they've had a long interest in it too. And now we go forward into the clinic. Oh, please let the drugs work in diseases! I can't wait to see the results of the trials.

Finished the day with two more events. The charity GOAL wanted to give their staff an update on COVID-19. GOAL will be involved in the vaccine roll-out to the developing world. And then in the evening, a lively Zoom call with the UCC Student Society.

Went to bed in a daze – and slept like a top.

FRIDAY 6 NOVEMBER

Morning discussion with the graduate students in Athlone Institute of Technology. Then a lab meeting. Tristram presented. Excitement! He has seen that our wonderful (ha!) potential new drug itaconate might block coagulation in vivo and could well have the makings of a new medicine. This could be big. We are aiming for COVID-19, as the evidence that this disease is caused by blood clotting in the lungs gets stronger all the time. Let's see what happens next.

We also found that a related drug called dimethyl fumarate (DMF), which is already used in humans to treat the inflammation in multiple sclerosis, might work too. Another example of drug repurposing? It's happening all the time in COVID-19 trials. There are efforts to block the cytokine IL-6 with a drug that is used for rheumatoid arthritis. So why not a drug for MS being redeployed against COVID-19? Early days, don't get too excited. But Tris's data got us all going, and fair play to him.

MONDAY 9 NOVEMBER

Top of the agenda with Pat today was how long immune memory might last against COVID-19. Then onto mink, and how they might have been infected. Of all the creatures … This is leading to culling of millions of mink – such a shame. A friend of mine with dark humour sent me a message: 'Satan is summoning the mink!' And an analysis of mental health issues on Twitter has shown a 14 per cent increase in people looking for help – from 40 million in 2019 up to 60 million in 2020. So tough for so many.

And then – pure magic. Magic that might change everything! Pfizer/BioNTech issued a press release saying their vaccine trial had revealed that the vaccine is 90 per cent efficacious. I nearly fell off my seat. This is what we've been waiting for. The number 90 becomes the best number ever. There was always a risk that the vaccine would not work or might have some side effect. I really thought it would take until March or April to reach this point and yet here we are!

It's an RNA vaccine, which is brand-new technology. An RNA vaccine has never been approved before. And now we have efficacy, and at a remarkable level. A flurry of media uptake. Rory came to the lab to film for Newstalk. *Drivetime* and *Six One* want a slot. *Claire Byrne Live* also. Such excitement. What bothers me slightly though is we haven't seen any data. It's science by press release. But still, huge optimism. This is definitely the beginning of the end. There's a good chance that the Moderna vaccine will work, as that is also an RNA vaccine. A number of vaccines, including AstraZeneca/Oxford and Johnson & Johnson, are DNA and not RNA vaccines, but they also make the spike protein, like the Pfizer vaccine. The immune response to the spike protein, coming from RNA or DNA, is strong.

Looked up a bit of the history for the media. BioNTech are a German company founded by Özlem Türeci and Uğur Şahin. Şahin is a medical doctor with a PhD in immunology. He had worked in the lab of Rolf Zinkernagel, who won the Nobel Prize for his work on immunology in 1996. Türeci is his wife and the Chief Medical Officer of BioNTech. They have been pursuing RNA vaccines for years, initially against cancer. They began making the vaccine back in February, and now, nine months later, the world has a vaccine that is 90 per cent efficacious against COVID-19. A number of things were needed. First the RNA had to be modified to make it safe, and that was done using technology discovered in the University of Pennsylvania by Katalin Karikó. And importantly, a way to deliver it, which involves liposomes, made from specific fat molecules. That had to be invented too. This was achieved in the

University of British Columbia. If ever you needed evidence for the importance of universities, this is it. An awful lot of science had to be done to get to this vaccine – and in 20 years or more.

RTÉ *Six One* came in to interview me live in the Explorium – a big science centre where I was already doing some filming for the RTÉ science show *Future Island*. Great to be able to tell people – a real hope at last. I hope I conveyed the excitement, which is actually difficult to contain! I'm still smiling. I really thought it would be March before we got to this point, and even that was uncertain, with maybe low efficacy.

I feel like a kid on Christmas morning. I sense a kind of magic in the air. The colour of life will change from now on, away from black and white and eventually back into brilliant Technicolor. A long way to go, and many bumps likely, but there is hope …

TUESDAY 10 NOVEMBER

Zoom call with the Association of Community and Comprehensive Schools: 96 schools are in the group, and a load of principals and deputies. It's been tough for school management. Gave them an update and banged on and on about the Pfizer/BioNTech vaccine. Hope I didn't bore them.

Then out to the Explorium again. We go live tonight!

I have a lovely room of my own there with a nice heater. Essential, as today was so murky and cold. There's catering too, so every so often I get brought a nice cup of coffee or some lunch or dinner. I could get used to this.

Liz Bonnin and I did a rehearsal of the opening section and the big parts in tonight's show. We changed it a bit, leaving some topics out and adding in others. We also worked on a maths piece on the Fibonacci sequence. This is a sequence of numbers that can generate a ratio seen throughout nature. Did we invent maths or see something that was already there? Towards the end, I quipped, 'So this guy Fibonacci and all his numbers. Was he the original Count in *Sesame Street*?' I was proud of this line – the maths was tricky enough!

And then we went live. Eek! But it was fine. Interviewed P.J. Gallagher, who had driven a driverless car. I pointed out that I couldn't drive. 'Well,' he replied, 'at least that's something I can do better than you!'

Had to stay late as I took part in the COVID Citizens' Assembly, which is a real-time online event sponsored by SFI and happening every night after each show. It went well, though a couple of anti-vaxxers sent in comments. If you were tired and cranky you might get cross with them, but I didn't. I did, though, manage to spill a full cup of coffee on my trousers, but that wasn't because of them. It was because I am an eejit sometimes.

At least five people said to me that either they or their relatives – usually their mothers – were fans. Good Lord.

Got home about ten and put the TV on. Watched some movie, can't even remember what, and fell asleep on the couch.

WEDNESDAY 11 NOVEMBER

Interview with the Irish Strategic Investment Fund at Treasury Dock, North Wall, about Inflazome. They are also beneficiaries from the deal, which means the Irish exchequer benefits too. I got off the Dart at Grand Canal Quay and walked over the Beckett Bridge. It was lashing rain and windy, so it was a struggle. My umbrella was pushed inside out as I crossed the Liffey. You know that kind of day when the weather is against you? A nice hot cup of coffee sorted me out.

Over to Trinity for an in-person lecture. Imagine! It was to the MSc students in immunotherapeutics. Now there's something relevant. About five of them were in the lecture room, all wearing masks. The rest on Zoom. I really enjoyed it. I love lecturing, and it has been such a long time.

Con sent me another poem today. Has he nothing else to be doing?! It's a really nice one – about a priest he met in St Finbarr's church:

But there is this fellow
Luke O'Neill
Seems to know
What to say
Even wrote a book
'Never mind the –'
Oh! but –
Can't say the title now
Available I'm sure
In Bantry bookshop
Bantry Bay

THURSDAY 12 NOVEMBER

Today is my 27th wedding anniversary. I always buy the gift of the year. I couldn't get to the shop, so Orla, who is helping me on *Future Island*, kindly said she'd get it for me – a piece of sculpture. Felt a bit guilty, like one of those guys in the 1950s who gets his secretary to buy a present to give to his wife. Luckily, Marg liked it. Strangely, opposite the entrance to Explorium is Lamb Doyle's, the pub in which we had our wedding reception. Time slip. It's a sad-looking place now, though – all shut up.

Yet more vaccine news. Bring it on. Sputnik V is reporting 92 per cent efficacy. In your face, Pfizer. They have made 1.2 billion doses and will give 500 million to countries outside Russia. Two vaccines with high efficacy now.

Then out to Explorium. Liz and I talked about biomimicry, which is so cool – learning from nature to make something more efficient. Liz's passion is the Earth. We talked about the kingfisher's beak, and how its design influenced that of the Japanese bullet train as it decreases noise and is more aerodynamic. Whale fins have bumps on them making them more efficient as they glide through the water, and that design is being used in wind turbines. Incredible, really, the world we live in.

Almost half the show had been pre-recorded, and we pre-recorded the wrap comments too. Usually there would be some kind of a party, but because of COVID-19 this wasn't possible. But we did manage a big group photo.

We all said our goodbyes. I gave Orla an advent calendar as thanks, and told her that every day she opens is a day closer to the vaccine. I gave Liz a copy of *Never Mind* and a Trinity College Dublin hoodie. As she's a graduate I thought she'd like it. I admitted it had recently been given to me as a gift but was too small. Ah, the joys of regifting.

We bumped elbows goodbye and got into our separate taxis. We all said 'Here's to the next time!' It was striking how much effort went into these shows. All

I had to do really was turn up and do what Orla told me. It's much more involved than radio. Liz said the same thing. Radio is more straightforward. You kind of have to learn your lines more for TV. I wonder why? Is it because you're more self-conscious on TV?

Got home and we had a take-out dinner that Marg ordered for us. Anniversary during lockdown. Looked at our wedding album, which we sometimes do on our anniversary. Friends and family with love enough enough to light the room, and all so young.

FRIDAY 13 NOVEMBER

Went up to the Long Room shop today to sign some copies of *Never Mind*. Trinity's most iconic building. Mothballed since March. Book of Kells closed. Normally there would be almost one million tourists per year. And in 2021? Zero. Really brought it home to me how museums and visitor attractions have died a death this year. I left the shop with some melancholy in my soul. An empty shop in the run-up to Christmas with most of the lights out, when normally it would be a bustling place, with people buying gifts. Ghost of Christmas past.

It was getting dark as I walked through Front Square and it was only 4 p.m. The yellow lamps reflecting in the wet cobbles.

Into *The Late Late Show* again. Third appearance. Sure, I'm a regular now. Met Ryan in the almost empty green room. He told me how they had recorded some good stuff for the toy show, recreated 'Singing in the Rain'. It was going to be very special. We talked about the vaccine on air to the usual empty studio. There are now 1.2 million deaths, but Ireland has done well – third-lowest per capita in Europe. No mean feat.

We also did another demo using a special fluorescent dye on how to wash your hands, just to remind people of how important the simple things are. Coffee after in the green room with the crew. There was a feeling that we're all in this together.

MONDAY 16 NOVEMBER

Had trouble getting up this morning. The darkness outside and knowing there will be many more weeks of all this. Some mornings are like that. I wonder is it because of intense days and my brain going, *Ah here, hold on*? But after a scalding hot shower and cup of coffee the gloom lifts.

On with Pat and a vaccine update. This will happen every time I'm on now, as it's the main game in town. Also spoke about Thanksgiving in Canada, on 12 October, which resulted in a big surge in cases because of all the travel and social mixing. A warning for us as Christmas approaches. But will anyone listen?

Then did a piece for RTÉ (again!) for the Dublin Book Festival. Then I headed out to the Helix in DCU for more TV stuff, this time on a show called *Clear History*. This is a kind of quiz show where we get to expunge things from history, including your own! Kevin McGahern of *Republic of Telly* fame is the quizmaster. Colin Murphy and Joanne McNally are team captains.

I was in the dressing room just before filming and news came in that the Moderna vaccine trials results had shown even greater efficacy than Pfizer. Another Monday, another spectacular vaccine result. RNA vaccines really work.

The filming was some craic! I got to cancel from my history the time I was rescued off Rosslare by the RNLI. I was on my way to Cork in my then boat when the engine failed. The biggest RNLI boat in the fleet came out to rescue us. It turns out they were making a documentary on the RNLI at the time, so everything was filmed. They even interviewed Marg, who had stayed ashore, asking her if she was anxious. She said not really! And in the *Examiner* the next day the headline was 'Boffin rescued at sea'. In the end it didn't even make the cut in the documentary as it was all a bit boring. They all seemed to like my story. Especially the part when I set off a hand flare and it blew up in my hand, burning me.

TUESDAY 17 NOVEMBER

Didn't sleep well last night. Overthinking everything. Remembering what I was doing this time last year – giving a talk at Northwestern University in Evanston, Illinois, and also speaking at an event organised by the Irish community in Chicago. That was great fun – I even got to meet the guy who organises to have the river dyed green on St Patrick's Day! And then went on to San Francisco for a mini-sabbatical in Stanford. I had rented an apartment with Stevie, who was doing a project in Stanford as part of his MSc, so got to spend time with him too. It's where I wrote a few chapters of the book. I had the TV on in the background as I wrote – the Star Trek Channel. No chance to go to the US now. Not with Stevie either. Got up around 3 a.m. and went downstairs to make myself a cup of tea. Then back to bed. Finally nodded off.

On the agenda today is a discussion with the Fianna Fáil parliamentary party. Lots of questions. Many about whether the damage being done to the economy was justified. Did my best to answer with the usual mantra: it's not about wealth or health, it's about wealth via health.

I was asked a pointed question about whether Dr Martin Feeley (clinical director of the Dublin Midlands Hospital Group) should have resigned after he had made comments about COVID-19 being no worse than flu and that people at low risk of the virus should be allowed to be exposed to it, which would enable the country to develop herd immunity. I had signed a letter to *The Irish Times* questioning what he said. The HSE rejected his opinion. He went on to resign his post. I said this was regrettable, and that he shouldn't have resigned, but instead should have defended what he said with data. Where's the data to say it's no worse than flu? Because all the data I've seen says it is worse, with a significantly higher mortality rate for people over 40. And how many people does he think would die or be afflicted with long COVID if we allow low-risk people to be exposed? The data mantra yet again. So important. All that should count when it comes to statements on COVID-19 is data, especially if there is a dispute. Defend your conclusions with data when you can.

Following the parliamentary party discussion I got another magical phone call, this time from Sarah at Gill to tell me I'd won the Best Popular Non-Fiction award. WTF! Keelin Shanley's book *A Light That Never Goes Out* was in the same category. I hope hers wins the other category she is up for – I'll feel bad if it doesn't. But it's a good feeling indeed. A science book winning this award! Times we live in, I guess.

Went home on the Dart on a high. Some day, and on three hours' sleep.

WEDNESDAY 18 NOVEMBER

Day started with an interview by Skype on Euronews. Also, on Spanish TV – Telecinco – with Cristian. And later with Kieran on the Hard Shoulder in Newstalk. He asked me had I seen the data yet on the Pfizer vaccine and I had to say I still hadn't – just the press release. Catherine also called me from WAKA TV, asking me if I'd be interested in doing an episode of *Keys to My Life*. This is where they bring you to houses from your life that matter to you and you talk about them. I said it would have to involve the UK, though, as I had six addresses there when I was a student in London and then a postdoc in Cambridge. We could start in Bray in my old home or maybe my grandmother's old home, which overlooked ours (Catherine suggested that, in case we can't film in my old home). I told her my first address in London was on the thirteenth floor of a tower block in Mile End. Then I moved to North London, then East Sheen (much posher). I said we could film in the Royal College of Surgeons, where I did my PhD. There's a wonderful museum there called the Hunterian Museum, which has a skeleton of the Irish giant. They called me that when I was there because of my height. Then we could go to Cambridge. Great places to film there too. Then back to Ireland. She said it would be great! But of course, it would have to wait until

COVID-19 has passed, so I'm thinking, this will never happen. Or when it does, they won't want me because I will have faded back into my life as a scientist … which wouldn't be the worst. I'd love to do it, though, for the sheer fun of visiting those old haunts of mine.

THURSDAY 19 NOVEMBER

A few emails worth recording:

> I am a school principal and I haven't slept in over a week because my teachers are refusing to run Christmas exams because they are scared of becoming infected. Can you please reassure us that we won't pick up infections from exam scripts?

> ....

I was happy to reassure her.

> Thank you very much for all you are doing. I know it comes at no small cost to yourself, your personal life and your energy.

> ....

This person must have seen me this morning! Nice to read this supportive email.

> I had a dream about you, Luke. You were in a field with long grass, wearing a white lab coat. I woke

with a feeling of well-being. Thank you, professor, you help me so much with my anxiety.

....

No haters, but then I block them. A friend sent me a message they had seen on WhatsApp: *O'Neill is in the pocket of big pharmaceutical companies. He has a conflict of interest when talking about vaccines and therefore no credibility.*

Someone put a reply under that: *What credibility do you have, ya plonker?*

And a couple I can only assume are slags:

The only people I can think of who carry similar on-screen presence are Jimmy Hendrix, Bruce Lee, Arnold Schwarzenegger and perhaps David Attenborough.

....

My friends are asking me to ask you if you would replace the baby Jesus in our nativity crib because you're the saviour of the world.

....

That one came from my dear friend Betty Ashe, who has headed up the Pearse St residents' association for years. She knew I would love it!

On with Pat this morning. Discussion on vaccine

uptake and what the possible level of vaccine hesitancy will be in Ireland. Someone texted in to say I should be the next Minister for Health as I don't waffle – I pointed out that would disqualify me from the job. Also discussed how the 'proning' of patients (putting them on their stomachs) is having a big effect by allowing fluids to drain from their lungs, easing their breathing. Something as simple as that was first done in 1976 by ICU nurse Margaret Piehl and Dr Robert Brown. It's now a standard procedure and is saving lives.

Another in-person lecture to the MSc students on the story of Inflazome. They seemed a bit nonplussed, to be honest, which just shows just because I think it's fascinating doesn't mean everyone else will. I've seen that before. Sometimes a lecture really works when you least expect it. It reminded me of a scene in one of my favourite movies – *School of Rock*. Jack Black has put together a band of school kids and to inspire them he gives them CDs. He gives Zack the guitarist a CD by Jimi Hendrix. Jack is 'psyched' giving it to him but Zack just shrugs. A look of disappointment comes over Jack's face, which is exactly what I had at the end of the lecture when there was a lukewarm reaction. Maybe I misread it. But hey, such is the life of the lecturer.

Then a Zoom call with a school in Waterford. Lots of questions. One guy asked whether he will be able to go to his debs. Only if he gets vaccinated was my reply. I mentioned how there's talk of people having bracelets to show they've been vaccinated and how that will allow people into pubs and theatres. People could wear it like a badge of honour.

Had a 45-minute masks-on tutorial with three students. The topic: sex differences in disease susceptibility. This is a fascinating topic in immunology. Women have a much higher risk of some autoimmune diseases (for example, lupus), while men are at higher risk of COVID-19, and we don't know why. It looks like there are big differences in the immune systems of women and men and it's a very hot area of research.

Finished the day with coffee in the Provost's house. Paddy Prendergast wanted to congratulate me on the sale of Inflazome and to discuss what Trinity might do with the money from the deal. I said it would be good to use it to cover the costs of staff and students who might want to go on courses on how to be entrepreneurs. Or maybe even to invest in new Trinity spin-out companies. Put the money back into the system to help others. Paddy liked the idea. I'm so happy that Trinity has made some money too. Trinity was always very helpful to me in my dealings with companies.

It felt very dark on the way home tonight. Dart was almost empty. Dark and cold walking home. It sent a chill down my spine. I can't help but feel the start of a tough winter. Respiratory viruses like COVID-19 love the winter – they spread indoors. It's not going to be easy.

Those thoughts inspired me to write a piece for the *Sunday Independent* on what Christmas might look like during COVID-19. I urged caution. Good ventilation in houses essential. But will anyone listen?

FRIDAY 20 NOVEMBER
Recorded a Zoom call with Gary for the Irish Book Awards. I had to act surprised and say a big thank you!

Took part in a conference on COVID-19 organised by my colleague Seamus Donnelly for the Association of Physicians. It was themed 'The COVID Enigma'. Garret FitzGerald spoke too, from Pennsylvania – always the voice of reason and one of the best physician scientists ever to come out of Ireland. It was a great session. All either scientists or physicians, which made a nice change from my usual engagements on COVID-19. No one was especially optimistic, it has to be said. Or at least in the short term. There was clearly much joy over the vaccines, but a long road ahead.

Went for a few drinks with Colm down beside Sandycove beach. It was chilly tonight. But Colm brought a flask of hot water and some whiskey, which helped. He's working as hard as ever in Holles Street. I often think how tough his job with neonates must be. If he gets a call, he knows a newborn baby is in trouble. We talked music too – wondered if The Metabollix will play a gig again. We got high on the whiskey and chat. The twinkling lights towards Dún Laoghaire looked so pretty ('red lights, green lights, strawberry wine') and the tide slowly came in over the sand. If only it wasn't Baltic cold. We then went round to Brian's garden for beers. These physicians certainly know how to drink. I'm always amazed at how in conferences it's always the liver doctors propping up the bar.

He slagged me off for all the name-dropping I've been doing. I said, it's funny you should say that. The other day I was on a Zoom call with Claire Byrne, Ryan Tubridy, Miriam O'Callaghan and Pat Kenny …

SUNDAY 22 NOVEMBER

I stayed in bed yesterday until 3 p.m. All snug. No demands.

The controversy over RTÉ's top presenters taking part in a retirement party erupted today. Lots of apologies: Miriam O'Callaghan, Bryan Dobson. Photographs were taken. I wonder who leaked them? Symptomatic of the times we live in, I guess. Such a relatively minor event that was probably low risk. And they get a kicking for it. Like the 'Golfgate' incident in the summer when a bunch of politicians and friends played at a big golf outing and had a dinner afterwards, in breach of the rules. This led to EU Minister Phil Hogan resigning. It took him a few days, but he eventually went. Everyone is being watched all the time.

Got a nice email today from a daughter of one of my old professors in Trinity, Cyril Delaney. She thanked me for all I'm doing and said Cyril would be proud. He passed away a few years back. He gave us our first physics lectures in the old physics lecture theatre.

But that lecture theatre is much more famous for something else. It's where Erwin Schrödinger gave his 'What Is Life?' lectures, and where we filmed a documentary recreating the lectures. Those Schrödinger lectures were very important, as they were the first to consider what genes might actually be made of. The book based on the lectures, called *What Is Life?* was hugely influential.

But that lecture theatre is also important to me for another reason. I was called out from one of Cyril's lectures in that theatre on 25 February 1982 for an urgent phone call from my Aunt Stella telling me to get to St Michael's Hospital in Dún Laoghaire as quickly as I could because my mother was dying. She died that night. It's interesting how we humans put huge relevance on places. For obvious reasons, I'll never forget the large physics lecture theatre in Trinity.

MONDAY 23 NOVEMBER

Opened the Trinity College Awareness week on Zoom. Minister for Higher Education Simon Harris was on too – we seem to be doing lots together. I had to do this from the Newstalk studios, as I was on with Pat immediately after. Split-second stuff. More good vaccine data today, hooray! The Oxford/Astra Zeneca vaccine. Yet again, great efficacy, but it's a bit confusing. If they give a low dose followed by a high dose, they get 90 per cent. But two high doses give 70 per cent. Not clear why but hey, it's all good. Waiting for all the data. Imagine if there were several different vaccines in the first few months of next year? That will lift everyone and change everything, although if I see the words 'game-changer' once again, so help me ...

Eimear in Newstalk said there was a fair bit of trolling going on, so that slightly perturbed me, but I'm getting used to it. Got home and felt sick of COVID – not an unusual response, I suppose. There are times when even something you're really interested in pales. Like when John Lennon sang about the times when even he hated rock and roll.

I sent a supportive message to Danny Boy, a guy on Twitter doing a fantastic job of informing people on the COVID-19 situation. He has been viciously and repeatedly attacked for it, and on one of his blogs he seemed very down. I told him to block the haters. What the hell is wrong with people? The ones who attack are either scared, in denial, or are in it for political or financial gain. Nastiness holds back the truth, that's for sure. Anyway, I hope my words of support helped him.

TUESDAY 24 NOVEMBER

Bit of hate came my way today too. The Zoom call I did for the Waterford school was filmed. I didn't know that. And then it got shared around and my comments on vaccination and the wearing of a bracelet to show you've been vaccinated, which is being discussed in some countries, has triggered hate. *Who are you to insist that children be vaccinated? And have to wear a bracelet, like being branded?* Didn't see it myself, but was sent text from Twitter. Deleted it. Also, someone tampered with my Wikipedia page, maligning me. Got Sam to change it back. But today was actually a good day because I was on CNN. Yep. Put that in your pipes, haters. Discussed vaccine roll-out for poorer countries.

Ireland moving to Level 5 to save Christmas. The government have said that if we restrict our movements over the coming weeks, it can loosen restrictions slightly for Christmas. Is it in any way feasible? Is it because they know we will gather at Christmas anyway, and that they want to help make it as safe as possible? The plan seems to be focused on getting the viral case numbers right down over the next couple of weeks and we'll relax things a little. Seems misguided, and will surely cause a spike in January, as many are saying. It might not, of course, but there is a risk.

Was on the Claire Byrne radio show. She wanted to chat about what I had put in my last *SINDO* piece, on how to have a safe Christmas. I said ventilation was key for Christmas Day. I made a clumsy joke, saying to stick Grandad by the open window. To limit the number of people. I also said how the CDC recommended various things, like to bring your own dinnerware if you're visiting someone's house. Claire brought up the CDC guidelines and said the instruction was not to pass around the gravy boat.

On *The Tonight Show* too with Matt. Ida Milne was on with me, and it was really good. We chatted about vaccines and the history of pandemics. I like being on with Ida. Another decent person I've met during these strange days.

The talk in the media today was about loosening of restrictions for Christmas. Non-essential retail and restaurants might reopen. I can't but feel we are in for a rocky time. Will try to tell people what to do to avoid an upsurge but, as Tony Holohan said in the summer about travel, I am beyond nervous.

THURSDAY 26 NOVEMBER

Today on Newstalk we spoke about a study on which musical instruments are the worst for spreading COVID-19. The things scientists study. Turns out to be clarinets and oboes – the reed seems to generate aerosols, which carry the virus.

Came out of the studio and a couple approached me. The man said he was with *The Irish Inquiry*. Shoved a microphone in my face. The woman began filming. I said I didn't want to be filmed. He said his readers had questions for me and I hadn't answered the email requests for an interview. I hadn't received any, I said. I kept walking. He began asking me questions. Did I have any shares in drug companies? Was I not conflicted because of my connection to those companies? I told him I didn't have any shares, which seemed to surprise him. Vaccine questions too – safety issues. And why wouldn't I have a debate with Dolores Cahill? I said, well, what normally happens in scientific discussions is they are hosted by learned societies like the RDS or the Royal Irish Academy. I said that if they organised it, I might. Got back to the lab and told everyone and they were worried for me! What if he'd become violent? Someone said he would cut the interview to twist my words. I tried to remember exactly what I'd said. We'll see.

FRIDAY 27 NOVEMBER

Got some emails about my interview on *Claire Byrne*. The thought that I might be denying people a good Christmas has riled some. A man emailed to say that I seemed to be losing my mind, recommending that people should put their grandparents beside an open window. I'd made that comment to underline the importance of indoor ventilation. He said he'd seen it before: someone comes into loads of money and then loses their mind. He said he could recommend a good psychiatrist, but not one in Dublin, where I am too well known. Initially I thought, the cheek! But then read it again and realised he was trying to be funny.

Waiting for a call from a US collaborator and made the mistake of answering my office phone. Got a huge diatribe from someone shouting at me, saying I was a disgrace for insisting on the now-infamous Waterford school session that children be vaccinated against their will and be branded. I said I hadn't said that, but I suspect I wasn't heard.

I can't help wonder, am I doing too much of this? I can't believe the interest in science and medicine in the media, although I guess it's obvious why. I hope that young people might be inspired by listening to us scientists. They might want to become scientists themselves. Never have there been so many in the media.

SATURDAY 28 NOVEMBER

I see from my diary that I'm meant to be in Malawi at an immunology congress. Was really looking forward to going. I've hardly been to Africa (apart from South Africa and the infamous trip to Tanzania earlier this year). Given that there are so many infectious diseases still rampant on the continent I've always felt we should have more international conferences there.

Went to Noel's butchers today and ordered the turkey for Christmas. A ritual in which he always reminds me of the year I forgot to pick it up on Christmas Eve because I'd been waylaid in the pub. He had called me at 4.30 p.m., asking me where the hell I was.

SUNDAY 29 NOVEMBER

The government has announced that the country will move to Level 3 on 1 December. We've all been good little boys and girls and Santa wants to reward us. Hairdressers will reopen. Restaurants too. No mixing of households though, with a slight easing around Christmas allowing three households to mix from 18 December to 6 January. Beyond nervous. Do we really think there will be hardly any mixing?

MONDAY 30 NOVEMBER

On Newstalk I spoke about sniffer dogs detecting

COVID-19 – yet more evidence for that. Dogs have a hugely sensitive sense of smell, and it turns out they can detect people who are positive for COVID-19. They are being used in some airports and, who knows, that might happen here.

Had a call with the CEO of AIB and the CEO of IBEC! An update on COVID-19 and what might happen next. I was unable to tell them, other than saying the vaccines will get us out of this.

A crew came in to film me for the RTÉ programme that will be broadcast for the overall winner of the Book of the Year. Good interview on science, writing and of course – yep, you've guessed it – COVID-19.

And so November ends. The Irish Book Awards. And we have vaccines! A slight lifting of restrictions. Perhaps life will be a little bit easier now. Just saying that makes me feel worried about what's to come.

DECEMBER 2020

TUESDAY 1 DECEMBER

Last month of a year like no other. Festive season to come. This virus doesn't care.

WEDNESDAY 2 DECEMBER

Thinking about the meeting way back in January with the SARS experts who said this was all going to be bad. If only we knew how bad. Somehow back then we thought it might be containable. It might be like SARS itself. A big deal for the countries that were hit, but not a global pandemic. It seems like so long ago, and yet also like yesterday. I can't believe it's 11 months since that Rotterdam meeting in that foggy, atmospheric place.

This all put me into a bit of a downer. I thought about all the people whose lives have been so affected by COVID-19. I thought about Desiree. She hasn't been out of the nursing home since March. And she mightn't have too many years left to her. Such precious years. We have to hope the vaccine liberates her and all the people like her. I wonder what the policy will be on that? Will they let the people out of nursing homes if they are all vaccinated? I certainly hope so. I guess the viral count in the community will have to come down substantially. They are vaccinated so they are at low risk of becoming ill. And then when they go back in, everyone else is vaccinated too. Surely that will be the humane thing to do. Sadness when I think of Desiree now.

But then I had a lift. The MHRA in the UK has approved the Pfizer/BioNTech vaccine. First regulator to do that. This means they have scrutinised everything and they are happy with what they have seen. This is a very good development! The European Medicines Agency can't be too far behind. The first vaccine to be approved for COVID-19. The gloom lifts!

The day ended with my first meeting with Brian MacCraith and the rest of the vaccine advisory group. Brian is heading up the national vaccination strategy and has asked a group of immunologists to advise him occasionally. I am delighted to oblige. Brian gave us an update. We discussed how so far there is no evidence that the various vaccines protect against transmission of the virus to others. This is because although the vaccine protects people in their lungs, it might not protect them in their upper airways – that would require what is called a mucosal vaccine. There's still a chance that there will be some prevention of transmission as people might have a lower viral load overall. This is a work in progress. We discussed how this might affect the messaging, which will be all-important when it comes to the roll-out. I got the feeling that we are just taking the first step on a long journey, but at least we've taken that step.

THURSDAY 3 DECEMBER

Into Newstalk early to record the first podcast, yay! Pitch-dark and cold on the way up to the Dart. A sense that we are now entering the depths of winter. And with this damn pandemic raging.

I had recorded some podcasts with Jess, but Eoin will now help. They had asked me to do a podcast with Pat, but instead they had taken clips from the interviews with Pat and put them into a separate podcast. But this is my own podcast. Gulp! I've always been keen to do it, and Sam has always said I should. I have to say I don't listen to many podcasts, but I will now. And I've done some with Blindboy, David McWilliams, Stefanie Preissner and Doireann Garrihy. They've always been fun. I thought about doing one off my own bat, but this is much, much better. Eoin is a great help and looks after all the tech. And Newstalk will plug it.

I chose *The Science of Love* as the first one. Love is all you need. Me and Jess had done it before, but now with Eoin on board, he will add music and sounds. It's all about the biochemistry of love – pheromones, oxytocin, dopamine. Also, the evidence for a broken heart. How people often die soon after a loved one dies, because of pressure on the heart. I did it in one take and really enjoyed it. Not COVID-19-related, which is a huge relief. And no uncertainties. With COVID-19 there's always the risk of getting something wrong, or maybe the science being corrected in the future. But with this, I'm using 'old' science, which is as true as it can be. So that brings satisfaction. As does the fact that I'm spreading the science word – the key mission in all this. A good feeling.

On with Pat after. Still strange that I'm in the studio and he's at home in Dalkey. Haven't seen him since March. What will it be like when we meet up again? We asked what gave listeners the most comfort during lockdown and music came out top. Yet again, COVID-19 teaches us what is actually important in our lives.

Told Rory afterwards that the guy who doorstepped me has put the video up on YouTube. He had a look and said it was nothing to worry about. If I asked to take it down it would just draw attention to it. I conveyed some of my emails of support: *You were doorstepped by a fucking looney and you kept it real. Fecking idiot doing that to you.* And: *I would have chinned him if I were you.*

Got back to the lab and met Alex for his PhD viva. Big day for him. Normally an external examiner would travel to Dublin. Alex would give a seminar in front of the whole department on his thesis and then get examined for hours by the external examiner as well as an internal one. The external examiner has to be a world expert in the area. We've asked Prof. Eicke Latz, one of the world's leading immunologists and the old friend I met in Nassau just before this whole thing started. It's an absolute ordeal for the student for obvious reasons, but it is an important rite of passage. It's a skill as a scientist to be able to stand up and defend your work. Alex seemed calm. I went with him to the room, and sat socially distanced from him with my mask on. He gave his talk on Zoom. It was good that his parents and his grandfather could attend. After his talk ended, the examination began.

I went back to my office and waited. It's a strange feeling. Like being an expectant dad. I really felt for him. I noticed that he was wearing socks with little footballs on them. He's a big football fan so I guess they brought him comfort! I did some work and wondered how he was getting on. After about two hours, the word spread to the lab that he was out! He came down to the lab and said it all went well. Eicke had asked lots of questions, but he could handle them all. Everyone was delighted for him – they are all so fond of him.

The lab had put up some bunting. After four years of hard work, he's now Dr Hooftmann. Huzzah! This is a great part of the job. We couldn't go out for dinner as we would normally do. When this ends we will have an awful lot of catching up to do when it comes to celebrations.

FRIDAY 4 DECEMBER

Thousands of restaurants and gastropubs opened today after being closed for six weeks. I hope this will be OK but I have a sense of unease.

Had a Zoom call today with Dermot O'Callaghan, who is the chair of the body that represents wedding bands. Lots of people joined. He wanted me to remind everyone of all the restrictions, and also to give them something to look forward to. So difficult for all those musicians, who made a living from weddings and other functions. I did my best to bring them some hope.

SUNDAY 6 DECEMBER

Fergal from RTÉ News came over to the garden to film an interview about the Pfizer vaccine. He had been to the facility in City West, where the first delivery arrived today. Great excitement! He said he had never got so much footage of a freezer. It all brings such hope.

MONDAY 7 DECEMBER

Another dark morning. Stygian gloom.

TUESDAY 8 DECEMBER

Big *Prime Time* interview in the lab on the Pfizer vaccine. People are worried about its safety. And I'm thinking about the logistics of vaccinating everyone. There are indeed challenges ahead, but surely we can overcome them? Other vaccines are coming – Moderna, AstraZeneca, Johnson & Johnson and more. Hope in bottles. And the UK is the first country to embark on a mass vaccination campaign.

Another interview – this time about John Lennon. I seem to have become the John Lennon guy as well as the vaccine guy. He was shot 40 years ago today. I can still remember where I was when I heard about it. The JFK moment. I was 16, in sixth year in Pres Bray, and was coming down for breakfast that morning. Put RTÉ radio on. It was the third item. I couldn't believe it. It became the first item later that day. It felt like I had lost a friend. My dad had encouraged me to go to Liverpool the weekend after he was shot to a big meeting of fans. My mother said no!

WEDNESDAY 9 DECEMBER

I paid a huge amount in tax today, from my earnings from Inflazome. That's the way of things, and I don't mind doing it.

THURSDAY 10 DECEMBER

Another cold dark morning and the news is bad. Angela Merkel is telling Germans not to go to the usual Christmas markets. No waffle stands or mulled wine. We need to be equally vigilant with our Christmas here. The metal seats in the Dart station were wet yet again but I sat on them anyway. A wet arse makes me feel alive.

Feel like I have a cold coming on. Might this be COVID-19? Nah! Went home after lunch and had a wee nap. Better afterwards. Watched CNN as per. The FDA will approve the Pfizer vaccine very soon. Trump is wondering why they haven't, seeing as how the UK have. Every agency has its own way of doing things, I guess. They will definitely approve it. I have no doubt of that.

And we reached another awful milestone: 1.5 million

people have now died of COVID-19. That horrible little scrap of zombie nastiness did this to them.

FRIDAY 11 DECEMBER

Dinner tonight in Casper & Gambini's in Dún Laoghaire. Marg had booked it because we're desperate to go to a restaurant. The good news is the tables were well spaced. The bad news might be that there were crowds at the entrance, and also near the toilets. That unnerved me. I could almost see a miasma of virus over people's heads. I will say, though, it was so good to be served! Great to eat food that we hadn't cooked ourselves.

A woman came over to our table to thank me for all the information I'm sharing and to tell me how her son was doing biomedical sciences. I said he had picked the right career. Felt some peace tonight when I got home.

SUNDAY 13 DECEMBER

Good spread in the *Sunday Independent* – wrote about the three ghosts as a warning as we approach Christmas. I'm nothing if not seasonal. The first ghost is Germany, where Angela Merkel says Germans shouldn't socialise at Christmas, including in the famous Christmas markets. The second is the US, where numbers are sky high, and where they are seeing the consequences of too much socialising at Thanksgiving. And the third is Manaus in Brazil, where the virus is unrelenting. I wrote: 'We must look after each other over Christmas and in the months ahead. Isn't that what Christmas is actually all about?'

MONDAY 14 DECEMBER

Another Monday. Even darker this morning. Thick of winter. Up to the Dart at 7.50 a.m. to get into Newstalk. Not many about. Sky was murky – sun coming up meant that there was a line of brown sludge. And the metal seats at the Dart station were wet as ever.

Perhaps we could combine the Pfizer and Sputnik vaccines? Like when NASA and the Russian space agency collaborate. We got into a bit of a frenzy talking about these – Pat usually ends our chat when it's especially science-heavy with 'Science never sleeps!' and he's absolutely right.

Popped into the Stephen's Green centre on the way back to buy the secret Santa gift for Anne in the lab. Got her something with 'Best Mum' on it. She's done well this year – she is like a mother to us all! It's a big Christmas tradition in the lab – I put on a Santa hat and distribute the gifts, usually flinging them at people. There have been injuries. We managed to do it, socially distanced. The norm would then be to go somewhere for dinner but as ever, not this year. Lots had Christmas jumpers on. Making the best of it.

On the way home I sensed a downer come on me. Feel sorry for the lab – so hard on them but they are bearing up superbly and plugging away with the experiments. Also feel like everyone wants a piece of me with the huge number of emails and the endless media requests. And then I felt bad for everyone else as well. Come on, Luke! Don't you know you should never take the monkey off the other person's back? Instead, help them carry it.

Once home though I got a massive boost from the news that Stevie will be able to come home – my boy's coming home! Played one of my favourite songs, Elbow's 'Open Arms' and ooh, a wee tear! Much chirpier after music as per.

Angela Merkel made a grim announcement today. Germany must go into a hard lockdown over Christmas.

TUESDAY 15 DECEMBER

Huge meeting with Roche on COVID-19 testing. What's the best way to do it? What about antigen testing? Some heavy-hitters on from India, Japan, China, Brazil, US, Germany and Italy. Japan are planning on having the Olympics next summer and will possibly test everyone at the entry points with an antigen test. Imagine that. If that works it might be adopted for all sporting venues.

WEDNESDAY 16 DECEMBER

Today was good. Up to Bressie's studio to take part remotely in the Christmas GPs session. They had asked me to play a song and I thought, *hmmm, I'll bring The Metabollix!* Colm and Brian came along. Here is 'Merry COVID Christmas':

> Are you hanging up your stocking on the wall?
> COVID changes Christmastime for all,
> But hope, it springs eternal,
> We'll beat this virus yet,
> As long as Santa doesn't bring mink for a pet,
>
> So here it is COVID Christmas, everyone can have some fun,
> Look to the future now, it's only just begun.
>
> Are the anti-vaxxers talking loads of shite?
> Are the anti-maskers moving to the right?
> Do you go swimming in the sea?
> Wear a Dryrobe you have made?
> Or eat banana bread then you've been Slade?

What will Tony do when he hears that Leo hasn't
been listening at all?

Are you waiting for the vaccine to arrive?
Have you got the guts to take it if it's live?
Does your granny always tell you
That vaccines are the best?
They did for polio, smallpox and the rest.

Colm sang the verse on the anti-vaxxers. I said he can
take the stick. We ended with a flourish – to complete
silence.

THURSDAY 17 DECEMBER

Fascinating evidence that asthmatic people are protected
from severe COVID-19. This is unexpected, as any
lung disease was thought to put people at a greater risk
because COVID-19 attacks the lungs. It's not fully clear
why, but it could be due to asthmatics having a slightly
more active immune system in their airways, which
protects them. Got a good response when we discussed
it on Newstalk – there are many asthmatics out there.
Sadly, another lung disease called chronic obstructive
pulmonary disease puts people at a higher risk.

And the Moderna vaccine was approved by the FDA.

Hooray! Three vaccines approved, all before Christmas. Unbelievable, but excellent, news.

FRIDAY 18 DECEMBER

The numbers are worse. I feared they would be. People mixing and getting infected. The government announced today that the extra loosening from today will now end on 31 December.

Dinner in Urban Brewing pub with a small number of the Inflazome team. Could be our last for a while. We ordered really good wine and sure, why wouldn't we? Couldn't go mad of course because we were out in 90 minutes, although they let us have one more drink in an outdoor part with a canopy over our heads. We toasted Inflazome and our good fortune.

Went home with a huge sense of contentment. Two-week break now. I really like this night every year. Home from work. A year done. Christmas to look forward to. Lit a nice big fire and sat on the floor by it, stoking away at the turf. Such a feeling of ease. Love watching embers. Something primordial about it. Us humans love the hearth. *Níl aon tinteán …*

This has been some year. So happy with how well *Never Mind* is doing. And of course Inflazome. And huge reward at the media work, as I think it's helping people and I am able to express my love of science. Very lucky to be in a position to do it.

SATURDAY 19 DECEMBER

Pre-Christmas lunch in Bresson in Monkstown with friends. We sat in an outside area and the wind kept blowing through the awning. Great to be served good food and to drink some wine. Saw crowds in Monkstown and Dún Laoghaire. Just gave me a sense of foreboding.

Got home and put up the tree. Love that part of Christmas! Lights, tinsel. I recommended in my *Sunday Independent* piece that everyone should put up decorations outside the house too, to brighten up the neighbourhoods. We put some lights around two trees in the garden. Takes away some of the gloom.

Our Christmas decorations go back over the years. I've got a lantern that I kept from my old home in Bray. Marg has a fairy that she has kept. And we have some decorations the lads made when they were in primary school. So the tree is always that bit sentimental, which is appropriate at Christmastime. A tree with memories and lights and warmth. The yearly battle with untangling the lights almost lead to a temper tantrum. Why is it we just jam them into a box every year when Christmas is over? I bet that is as good a personality test as any. Are you the kind of person who carefully stores your Christmas lights? Or are you like me, someone who just fecks them into a box? Once it was over, though, I could sit on the sofa, put the other lights out in the room and look at the wonder of it. Simple pleasures, maybe all the more vivid this year.

SUNDAY 20 DECEMBER

On with Brendan O'Connor and Stefanie Preissner. All the usual topics, but the big one was what will happen in January? There will be a spike. I'm sure of it. We have to hope it's not too severe, as the viral counts were low when we opened up somewhat. Brendan asked me about the New Year's Eve show on RTÉ One, which has been advertised. I've been invited to take part in a look-back and also sing with Mundy! He said that he would love to see me and Mundy being vaccinated on it. I said, bring it on!

Strange to think I'm going to be playing with Mundy in the Gaiety. Yet more oddness. Good of him to let me. I practised the song 'Mexico' a bit today too.

RTÉ News came over to film in the garden and it was broadcast on the *Six One* and the nine o'clock news. It's no wonder I'm getting slagged. Mario Rosenstock said the only show I wasn't on was the Angelus.

MONDAY 21 DECEMBER

Headed in to Newstalk: plan was to go in and then do a bit of Christmas shopping. Half-asleep heading up to the Dart and when I got there, I realised I'd left my iPhone at home. Like missing a limb. It had my notes. Headed back home for it. Ten minutes' walk at pace. It started lashing rain. Back to the Dart. Dart delayed. Panic stations. Rang Eimear to see if I could go on slightly later and she said sure, let's do it by Skype. Headed home again. Like a frigging yo-yo. Days like these.

It was fine in the end. Spoke about the new variant that has appeared in the UK called the Kent variant, or B1.1.7. Not much known about it. This is all we need. As happens often these days, they took a clip of it and had it on the news at 10 a.m.

I planned to work from home for the rest of the day, Christmas tree lights on. But John from *Claire Byrne Live* rang. He heard me talking about the new variant and said would I come on that evening? I said yes, fine. He asked me to retrieve the model of the virus, which is in my lab, so they could modify it to show what the new variant looks like. So, back up to the Dart again! Third time.

When I came home, Mundy rang to arrange a rehearsal. Mundy! I love 'July' because the lyrics remind me of St Stephen's Green. I started thinking about St Stephen's Green. I have one photo of my dad working on the deck chairs – I think it was taken in the early 1970s. One day a few months ago I took the photo with me to the Green and tried to find the spot it was taken in, using clues from the roofs that could be seen through the trees. I walked all around the Green twice but couldn't find the spot.

A car came for me and took me into *Claire Byrne Live*. Met Paddy Cole. Another hero. He has a book out too, so we had a chat about that. Did the interview on the new variant and the model looked great. Got home at 10.30 and someone texted to say Dustin the Turkey was slagging me. Said I wasn't a professor at all, actually a tyre-fitter. Fame at last! I liked one of Dustin's recent jokes: 'What's the difference between the Titanic and RTÉ? Titanic only had one orchestra.'

WEDNESDAY 23 DECEMBER

So now we know more about the new strain of SARS-CoV2 circulating in the UK. It doesn't make people sicker but it transmits more easily. A whole new worry. What if the vaccines don't work against it? Calm down, calm down.

Philip Nolan of NPHET said that the third wave of COVID-19 is clearly under way in Ireland. This will mean more hospitalisations, ICU admissions and deaths. Depressing and inevitable. We are therefore moving to Level 5 from tomorrow until 12 January at the earliest. Up until 26 December, though, three households can still meet up. No inter-county travel after that. Here's hoping for good ventilation in everyone's house.

Lit a fire out in the chiminea. Stacked it with logs, spitting and hissing. And made some hot ports for me and Marg. We sat there around the fire as the daylight faded and the Christmas lights came on in the garden. We probably wouldn't be doing such a thing if it wasn't for COVID-19, so there's a tiny benefit, I guess.

Paul Moynagh called in for me to sign four books. We chatted about our media experiences. I hadn't seen him since Galway, so I gave him the whole Inflazome story. He remembered how I'd strongly hinted at it. Went up to collect the turkey from Noel's butchers. In the queue for ages but time ran out as I had to go home for a call with Brian MacCraith and the vaccine advisory group. I thought I'd be able to get the turkey and get back for the call. Never thought the queueing would take so long, but it is Christmas I guess.

The call was excellent. Brian told us how there will be 10,000 doses of the Pfizer vaccine delivered on St Stephen's Day, with 40,000 per week through January. We all said we were there to help him in whatever way we could.

THURSDAY 24 DECEMBER

Christmas Eve in Newstalk. It's worse than working for Scrooge! Spoke about the positivity for the vaccination campaign that's about to begin. We played the recording of 'Merry COVID Christmas' by The Metabollix. Hope it gave people a lift.

Fair few people out and about. Got home and *phew* – I can relax at last! Watched – guess what? – *It's a Wonderful Life*. Well, you have to, haven't you? Even though the two lads are in their twenties they still wanted us to put up their Christmas stockings by the fire. I drew the line at leaving out mince pies and carrots.

FRIDAY 25 DECEMBER

Well, a Christmas Day like none on record for us, and for so many people. My sister is stuck in Brighton because of the recommendation not to travel from the UK. Swimming was banned in the Forty Foot to stop the crowds, which is another Christmas tradition we've had for 20 years. We went over to Coliemore Harbour, which was empty. A lovely morning – the three of them went in and I took photos. Dalkey Island a perfect backdrop. Yellow sand and the green island behind.

It was just the four of us for dinner. First time in, I'd say, 25 years that we've had Christmas without Desiree. We all mucked in. Went over to see her through the window after I put the giant bird in the oven. It was a nice enough visit. She had already had her dinner (but couldn't remember it). She was in good form. I gave her the book of old Irish photos coloured in. The one that knocked me off the number-one slot in the non-fiction hardback category. Not that I'm bitter – it's a wonderful book. I wonder how many groans there were this morning when people opened the wrapping to see my book. 'Ah! Not this fella again!'

Dinner was the usual. The best part is Stevie makes some special sausages with breadcrumbs. Everyone agrees it's the best part of the meal. I think, *Well now, isn't that typical.* I produce the turkey, ham, roast potatoes, carrots, sprouts, stuffing. And these are but naught next to some sausagemeat. I'm kidding, but you can't beat the family tradition of someone saying, as they do every year, 'This turkey is a bit dry.'

We played a game called *30 Seconds* after dinner. I never realised a board game could be a blood sport. Cliona sent me a picture of her having a drink in her neighbours' house. The neighbour is an old school friend of mine from Bray – Pat Ryan. And he has a bar in his house, with the bar top made from a bench from the science lab in our old school, Pres Bray. The picture showed him holding up *Never Mind* over the bar. Now there's spookiness. The bench where I first did science all those years ago. Another example of strange things happening.

And today's Christmas Day message from NPHET?

The new UK variant is in Ireland. Great. Couldn't they have waited until tomorrow? I feel this will be a theme in the coming months – new variants emerging. Why are they emerging as this virus mutates at only one tenth that of flu? It's because the virus is spreading so much. Every time it divides it rolls the dice and sometimes gets snake eyes. In the case of the UK variant, it is around 50 per cent more transmissible. Where before one person infected two, one person now infects three. R number is going in the wrong direction. Need to make sure we double down on masks and social distancing.

SATURDAY 26 DECEMBER

Sister-in-law Esme came over with her hubby Ciaran for dinner. This is another of our big traditions and has been going on for years. No kids with them this time. It was a lovely evening. Our single other household visiting us.

SUNDAY 27 DECEMBER

Good enough day. The traditional turkey curry. Had a piece in the *Sunday Independent*: 'Reasons to be cheerful'. Mainly the vaccine, of course. Watched some TV. I love that, when you just flick around and latch onto something and are taken with it. Must be to do with a relaxed state of mind. It was a documentary about *A Woman's Heart*, that wonderful album that we all bought, by Eleanor McEvoy and friends. I remembered how all those years ago me and Tony played support to Eleanor in Mother Redcaps pub, pre-Metabollix days.

We played *30 Seconds* again. More violence.

MONDAY 28 DECEMBER

Worked a bit more on 'Mexico' – I've learned it! Sam set up the mic and PA system and played keyboards. Can't get it out of my head now: 'Well if we go to Mexico …' Always great jamming with my lad. 'Open Arms' by Elbow. 'Let Me Roll It' by Paul McCartney. Brought me back to the Metabollix gigs. They seem so long ago and far away.

TUESDAY 29 DECEMBER

Up early and into Newstalk. I'm a glutton for punishment. Sure what else would I be doing? Dark and gloomy on the way in. Streets completely deserted. On Grafton Street a guy on a bike stopped and said, 'How's it going, Luke?' He said he was renovating the casino – the Georgian building in Marino. He said, 'You've helped me so much. I can't do much in return, but I'd like to invite you out and I'll give you a special tour.' It really touched me.

Good to see Tony on the Newstalk door again, and Eimear and Cormac. We're a community! Mark Cagney was there in person, which was great. Brian MacCraith was on first – interviewed via Skype about the vaccine. He said the first person to be vaccinated would be a grandmother from the Liberties. I felt a lump in my throat at the thought of it. Then I was on. We had a good general chat, mainly about vaccines.

Walked down to the Dart on an empty Grafton Street under clear blue skies. I can see clearly now, the vaccine's here. Oh God, bring it on.

Brian was right. On the news, the first person in Ireland to be vaccinated was Annie Lynch from the Liberties, who was vaccinated in St James's hospital. She, and the whole country, was delighted. First report of the virus was 12 January 2020 and a woman in Dublin is vaccinated 11 months later. The power of science. This all gives a real sense that there's a way out of this. It will take time, but there's a way out.

WEDNESDAY 30 DECEMBER

Over to Mundy's house for a practice. What a lovely fellow. Made us some coffee, then showed me his studio. He said it had been a tough year, but that he'd done a few Zoom gigs for corporates before Christmas, which helped. His wife Sarah came in and we had a good chat. The thrill of talking to people! I nearly needed a restraining order for the joy of it. We ran through the song a couple of times. We'll be grand.

An ominous speech from the Taoiseach. The country will now go to full Level 5 until at least 31 January. The nightmare has come true. He sounded like Churchill. So, it's clear now. All that mingling at Christmas has resulted in a huge increase in cases, much more than anticipated. We went out in early December, got infected and brought the virus into our homes over Christmas, infecting lots of other people. Thank God we have the vaccine coming.

Country back in lockdown. The third lockdown. Will we ever be free of it?

THURSDAY 31 DECEMBER

Well, a New Year's Eve like no other. Got the Dart into town, then walked up to the Gaiety carrying my guitar. A woman shouted, 'Ah Luke! I didn't know you could sing, as well! Is there any fucking thing you can't do?' I scurried along. In through the stage door, which is beside the Disney Shop on Grafton Street. For years I'd walked past it without even noticing it. They showed me to my dressing room, which had my name on the door. I said, 'I'm not going in there until there's a star on the door!' They had given me the boardroom of the Gaiety.

Lots of old posters on the wall advertising bygone days. Pantomimes. Peter Ustinov. Got me thinking about the history of the place. Bound to be a few ghosts around. Yet again, had to pinch myself. How the hell have I ended up here?

Tuned up my guitar and then went on stage for the rehearsal. Mundy there, with Sharon Shannon and her band. Great meeting her and all the guys. And the legendary sax player Richie Buckley. The band were great fun. Played the theme from the Pink Panther between rehearsals. We ran through the song. Sound was perfect. However, I got stuck on a line. Couldn't remember it. And ironically the line is 'Forgetting to remember.' Mundy was good about it and said write it out and stick it to the guitar. 'That's what I do.'

Rehearsal of the whole show. Chatted with Mary Coughlan and Brian Kennedy, who I'd met before on *The Six O'Clock Show*. We went to the green room (formerly a bar) and had a great old chat about the various diseases that afflict them.

A surreal evening then ensued. Mary was on first, so she was taken to the stage, and sang her first song. Then I was on a panel with Mary and Marty Morrissey, with Deirdre O'Kane in the chair. I said Tony Holohan was my hero for the year. Mary said Jon Bon Jovi, as he volunteered in a food bank in the US. I then went on with Mundy, Sharon and the band and we did the number. I remembered 'Forgetting to remember.' Mundy had picked 'Mexico' because he felt the lyrics were spot on: 'Promise me this will get better. And it will heal in the bright weather.' We sang it with gusto, bopping around.

Then back in the green room. As New Year's Eve approached Sharon played a bit of 'Auld Lang Syne'. We all sat around and watched the countdown and wished each other a socially distanced Happy New Year. No linking of arms and singing. No hugs or kisses. Just me, in a room with Mary Coughlan, Brian Kennedy, Sharon Shannon, Jerry Fish, Marty Morrissey, Jason Byrne, Sarah McInerney, Kathryn Thomas and Bernard O'Shea. WTF. All of us wondering what 2021 might bring. All of us a bit anxious but somehow hopeful. I didn't say how worried I was getting about the numbers of infections, which are growing day by day. Or about the UK variant.

I left around 12.30 a.m. on 1 January 2021. Stone-cold sober. Onto Grafton Street, which was completely deserted. No idea how I was going to get home. Walked around by St Stephen's Green. A drizzle started to fall, and I pulled my collar up around my neck. No taxis. Went onto the Free Now app, and amazingly got one. It arrived in five minutes. The taxi driver said I was his first passenger that night. He asked me where I'd been, and I told him. He said: 'Did you ever imagine a year ago that you'd be playing live on the stage of the Gaiety Theatre on New Year's Eve with Mundy and Sharon Shannon, with nobody in the audience?' I said of course I did. I'd had a dream of exactly that – except I was also naked.

This night gave me the exact moment that perfectly encapsulates for me this entire pandemic. I can't imagine anything as surreal happening to me again in my whole life.

And then I noticed a message on my phone. It was … the Taoiseach! He left a nice message to thank me for informing the people over the year in a clear and calm way and looked forward to engaging with me in 2021. This was so decent of him, and it made me feel so good. Couldn't quite believe it. Went to bed with my head buzzing.

JANUARY 2021

FRIDAY 1 JANUARY

The melancholy of a New Year. Especially this year. Stayed in bed late. Something struck me. We have all discovered a new way of hurting. Loved ones filled with fear and anxiety. Loved ones we can't hug. Loved ones stuck in nursing homes as their lives run down. Loved ones who can't come home. Loved ones who can't get on with their lives. Loved ones getting sick. Loved ones dying.

Shook myself out of those thoughts, and, like I do every New Year, I thought about goals for the year ahead. More COVID communication and research. Podcasts with Newstalk. Support the lab in their work. Teach. Hope for the trials being run by Roche. I'd love it if we made another significant discovery but this time for COVID-19. I have a great team now, so why shouldn't we?

Spent the day on the couch. It was a good thing to do.

SATURDAY 2 JANUARY

Went to buy some food with Stevie. Always good hanging with him, even if it's something as simple as buying groceries. Numbers of cases are climbing. Philip Nolan very gloomy on the radio. The nightmare is becoming a reality. What a start to 2021.

SUNDAY 3 JANUARY

Very bad today – 4,900 cases.

MONDAY 4 JANUARY

Back to the grindstone. It seems Christmas went by very quickly. Day started with an interview on Euronews at 6.15 a.m.!

Up to the Dart and into Newstalk. Streets empty and cold. On my way up Kildare Street I saw the cleaners I see every time I pass, on their smoke break. I wonder what they are feeling. I give out the information for them and everyone. They are welcome to listen. We did an update on vaccines. Israel would be the 'canary in the coalmine' – the indicator as to how well the vaccination campaign is doing in the real world. The UK is back into full lockdown, just like us.

TUESDAY 5 JANUARY

Sam came into work with me, as he wants to study in my office for his exams. Gives him a bit of discipline. Nice having him there. Headed out to the RDS for an interview with Eileen of the IDA for their annual staff event, all on Zoom.

WEDNESDAY 6 JANUARY

More non-COVID-19 work, hooray! Had a big planning meeting for a conference on the immune system and cancer. Such an important topic, as there have been great successes there with more to come. Then to a meeting of the Trinity COVID-19 centre. Ed Lavelle gave a super talk on vaccine adjuvants.

Lounging on the couch in the evening, I discovered a new TV show called *The Terror*. It's all about the Franklin expedition to find the Northwest Passage who got caught in the ice for two winters. It resonates – let's hope we only get caught in the ice of lockdown for one winter.

Took down the Christmas decorations because it's the Epiphany and that's what you're supposed to do, otherwise bad luck will befall you. Maybe last year the world didn't bother.

Number of cases now through the roof – thousands being reported every day. And the death rate is climbing. The big fear we all had has come to pass. But no one imagined it would be such a steep climb. The restrictions will turn this around, but when? How soon will we see progress? The government has announced all schools will stay closed until February at the earliest, as will all construction sites.

Switched over to CNN and saw a bunch of Trump supporters break into the Capitol building. Has the whole world gone mad?

THURSDAY 7 JANUARY

A little bit of snow this morning. Gave me a nice lift. Lovely to see the dusting on the roads with few cars. Now wouldn't it be great if there was a huge fall of snow, seeing as how everyone is at home? Days off anyway, with no need for the excuse of snow, but it would make everyone feel better.

In Newstalk I met Leo Varadkar. We had a good chat. He remembered me lecturing him when he was doing medicine. That made me feel old! I told him how the Taoiseach had called me, and he said, 'You'd never find me doing that!'

There are now robust predictions of what will happen once the vulnerable are vaccinated, and it's remarkable (but obvious). UK scientists are predicting a 90 per cent decrease in the death rate. That is the key metric. Herd immunity will be hard to achieve, but a decrease in death rate and hospitalisation rate will be huge, allowing the fear to dissipate. I will keep reminding people of that. The key role of vaccines, after all, is to prevent severe disease.

FRIDAY 8 JANUARY

On the news this morning – a further 8,248 cases reported (the highest for any single day since the pandemic started). Two weeks since Christmas Day – that's when many of the people testing positive would have caught the virus. And 2,327 deaths. Grim.

Nearly slipped on the way up to the Dart in the ice and snow. Big eejit.

Some more email messages over the past few weeks:

My dear professor,
We love your optimism here in Australia. We'd love to do a Zoom interview on what makes you optimistic.

From the Chief Optimism Officer, The Centre for Optimism, Melbourne, Australia.

....

You may remember I asked your advice on how to shoot a film. Well, we did it! Your tips on ventilation and social distancing, plus regular testing, meant we never had one case in the six weeks of shooting.

....

I've been feeding your words and information to teachers in Cambodia – thank you and hugs for the New Year.

....

Dear Professor Luke, I enjoyed your accordion-playing on the TV. I have two sons with autism. They are absolutely fascinated by vaccines. Can you please send them some information?'

....

SATURDAY 9 JANUARY

Another wonderful TV moment. I was sorting through the laundry with *Shrek* on in the background. It was a perfect combination of activities. The satisfaction of folding freshly laundered clothes while watching Shrek and Donkey.

But now the South African variant is here, as well as the UK variant. In the name of all that's holy, we don't need these variants. How to inform people without alarming them?

Today is the first anniversary of the first death from COVID-19, in Wuhan. One year on, millions infected, millions dead.

SUNDAY 10 JANUARY

Ireland now has the unenviable achievement of having the most precipitous growth rate in the world. *In the world*. And all because of Christmas. It's so upsetting. The bottom line is clear. This is a very contagious virus and all it took was for a percentage of us not to follow the guidelines – indoors, stuffy rooms, close contact. Bang – cases skyrocket.

MONDAY 11 JANUARY

Even though it was 9 a.m., it was dark and gloomy outside when we started the show. Pat asked me about a quote I'd given to Kitty Holland in *The Irish Times* – that there might be 100 deaths per day at worst. I said it was possible but that hopefully there wouldn't be too many days like that because the average age of those in hospital is lower and treatments are better. A lot of people texted in questions. Hope we gave some comfort.

The Taoiseach was on after me. The first question Pat asked him was, 'We have the worst numbers in the world. Are you and your government ashamed of that?' Today feels like maybe the darkest day so far. I took a step back and turned off the radio. Gathered my thoughts and reminded myself of the situation regarding the huge success with the vaccines.

Ireland had the highest number of cases in a single day per million of population of anywhere – I'll write that again – *anywhere*, in the world. There's real fear now that the hospitals will collapse from the strain.

TUESDAY 12 JANUARY

Watched another episode of *The Terror*. Fascinating. I'd read a wonderful book (*Erebus: The Story of a Ship*) by Michael Palin and saw an exhibition in Anchorage a couple of years ago. The show is pretty gruesome, though – attacks from a giant polar bear, people going crazy from tinned-food poisoning, most likely caused by lead. And cannibalism. Not quite as bad as Level 5 lockdown, so there's that.

WEDESDAY 13 JANUARY

Today started with a one-hour Zoom call with the whole of Mount Anville primary school. There were more than 200 attendees, the pupils and their parents. It was something to behold. It took Leah, the teacher who had invited me to join, at least 20 minutes to get everyone on mute. A cacophony! Finally, they got there.

She said later that every day was like that. I spoke for 15 minutes or so and then got loads of questions from all age groups, from six-year-olds up. They had watched my RTÉ Junior talk on the immune system. One question was 'What is your favourite memory as a scientist?' Another: 'What is the best discovery you've made?' I said my favourite memory was when I made my first discovery about the immune system. I told them my best discovery was of an important 'on switch' for the immune system that we'd discovered – we called it MAL. It was a joy to do.

THURSDAY 14 JANUARY

Recorded two podcasts with Jess. One on the ageing process and one on how the cold affects life on Earth, including winter swimming. Very topical. Really enjoyed them. She told me we'd got to number 3 in the Apple podcasts listings, whatever that is. I told everyone to jump in the sea.

Then a session with the Science Gallery on vaccines. They also had a clinical trials expert, someone who had had the vaccine and a psychologist talking about vaccine hesitancy. This was followed by a Zoom call with Frances, who heads up the Community Foundation. They had handled the millions raised on the *Late Late* toy show and will disburse it to those most in need.

Then did an interview with Arthur Beesley of the *Financial Times*. And RTÉ *Six One* came in to film. We talked about the likelihood of concerts coming back. *Prime Time* and *The Tonight Show* both asked me on tonight, but I couldn't because of everything else going on.

Realised I'm getting over-exposed. It's hard to say no, though, as I want to give people my knowledge. But I'm well aware of how annoying it can be for the same person to be on over and over again. We're all irritating, eventually. Our tics and turns of phrase. I keep saying 'gangbusters'.

FRIDAY 15 JANUARY
How badly is Ireland doing? Very badly. A lesson for the rest of Europe it seems. But numbers are climbing everywhere, and Ireland is not the only place where Christmas drove a huge spike. The key issue was the Irish love Christmas.

Got sent some more old photos today. Some of my old friends from Bray have been doing this. Me aged four at Tony Martin's birthday party. Me on a scout camp to Germany when I was 17, with 5th Wicklow Sea Scout troop. I sent that to an old Sea Scout pal, Jamie, who has been living in Holland for years. He sent me back a photo of him wearing his old '5th Wicklow'

T-shirt that we were all wearing in that photo. He'd kept it for all these years.

SATURDAY 16 JANUARY

This afternoon I had one of those magical, transcendent moments. In the simplest of ways. I sat down on the sofa having done the hoovering followed by the shopping. Put on the TV and *The Towering Inferno* was on. Pure escapism from my childhood. I remember seeing it in the cinema when I was around ten, with my dad. Steve McQueen as the fire chief and Paul Newman as the architect. It's just great to lose yourself in an old movie. I guess it's safe and nostalgic and somehow comforting. Even though in this case it's a disaster movie. Interesting that a blazing fire in a skyscraper is light relief.

SUNDAY 17 JANUARY

On Grafton Street today an anti-vaxxer grabbed the collar of my coat from behind and shouted abuse at me. I just walked on.

My piece on all the vaccines came out today in the *Sunday Independent*. A neighbour knocked and gave me a black cowboy hat, because I had said the spike protein is a bit like the black hat on the bad cowboy. The immune system is shown the black hat, and then when the bad guy turns up with the hat on, he's arrested.

Exactly how the vaccines against the spike protein work. He said that at last he understood!

I must be bored rigid as watched a football match – Liverpool v. Manchester United.

At 8 p.m. I did a live interview with ABC in Australia for a morning TV show. Beamed into a country where there is no lockdown! Feeling envious.

MONDAY 18 JANUARY

Wrote a piece with Christian on a recent paper showing that iron is essential for vaccines to work. They speculate in the paper on iron's role in COVID-19. Turns out if you're deficient, you might be in trouble, as your immune system won't work properly. It's by a lab in Oxford headed by Hal Drakesmith, who I know. I'll tell people to make sure their iron stores are up.

A kind man emailed me today. Amazingly, he's done the genealogy on my father's side and traced it back to Letterbrock, Drummin, Westport. We have a common ancestor in a family called the Berrys. Mary Berry is my great-grandmother. Her daughter Agnes was my grandmother, the one who emigrated to Salford when my dad was a baby. I'm a descendant of a Mary Berry! This must be why I'm such a good cake-maker. He also sent my grandfather's death cert. He died in Whitworth Hospital of liver cancer, which is what triggered my grandmother to emigrate to Salford. I knew that, but I though he had died of pneumonia. Getting all this information is an unexpected benefit of being on the telly!

TUESDAY 19 JANUARY

Had a big session today with HABIC, the Hair and Beauty Industry Confederation. Gave them an update on vaccines and various aspects of COVID-19 and when reopening might happen. It's been so hard on hairdressers and barbers.

THURSDAY 21 JANUARY

The new variants are causing concern, but let's see what the science tells us. Told our listeners all about mucus – scientists have used a method called Sinusoidal Outcome Tool 22 (SNOT22) to assess mucus from COVID-19 patients, and have found 375 proteins at a higher level than in normal mucus. Imagine that for a minute. Our snot is full of proteins. Gross joke coming up: maybe that's why we ate them as children. The fact that they have found ones only in snot from COVID-19 patients might prove useful one day. Also spoke about how losing your sense of smell and taste might actually predict less severe disease, which is unexpected. Not clear why this is.

Did a call with Deutsche Welle, a German public

broadcasting agency, which apparently reaches millions. I think what happened is journalists saw me on TV in Germany and then want me for their own shows. The good part of this is it never generates any hate. This might be what is called a control group. Why would an appearance on Irish TV generate hate, and yet on German TV not? Hmmm … ah yes, I've got it – good old Irish begrudgery.

After midnight I stayed up to watch the concert for Joe Biden's inauguration. Jon Bon Jovi sang 'Here Comes the Sun'. So appropriate. Oh I'd love to sing that one too, to an audience in Ireland, when we come out of all this.

FRIDAY 22 JANUARY

Watched *The Late Late Show*. A woman came on to talk about how her husband had died of COVID-19. No underlying condition. A complete surprise. It was heartbreaking to listen to it. Then a trad band came on, and the emotion in the music got to me. Lump in my throat. The power of music. My emotions are running ragged tonight.

SUNDAY 24 JANUARY

Heavy snow falling when I woke up. The magic of it! Brave Marg still went in for a swim. Window visit with Desiree. She was smiling away, as usual. I had the umbrella up – bitterly cold, sleet falling. Looking in the window at her smiling face. Left her a box of Lindor chocolates. One of the nurses had told me she scoffs those in no time.

Today was the first day of our three-course meal plan. My idea. One of us does a starter, the other a main, and the other dessert. Sam did crab claws, I did chicken cacciatore and Marg did a fruit salad. I wonder will we keep it up?

Case numbers still high, hospitals nearly full, government bickering and dithering. And these new variants – is there no way out?

MONDAY 25 JANUARY

Had a classic anxiety dream. I was due on Newstalk but was late and went into a park to find a quiet spot to do it by Skype. I dropped my iPhone into a flowerbed and it got covered in soil and wouldn't work. Panic stations! Couldn't call Eimear as the clock wound down. Felt really bad that I was letting them down. Woke in a sweat. Looked at my phone – working fine and only 6.30 a.m.! Had one of those great snoozes.

On the show I spoke about how allergy to the Pfizer vaccine is actually very rare. Only 21 cases out of 1.9 million people vaccinated. Hope this reassures.

Then I was on Al Jazeera! Yep. Something of a thrill,

it has to be said. Quick interview, again about Ireland's disastrous situation.

From 4 p.m. I was at the online Keystone conference on immunometabolism. It was supposed to be in Keystone, Colorado. Gave my talk. It was good! First online conference I've enjoyed. The science really resonated. Listened to four talks and they were interesting. Not the same, though. But nice to see familiar faces on the Zoom – my old scientific friends, still there. It's just so strange not to be going in person to conferences. A year since I did that. A year! Can't wait to meet them in real life again. It's the spontaneous conversations that really matter. Maybe something subconscious goes on, and then you get a scientific idea. Head has to be in the right place for it, and I think that happens more when you're actually with someone. At least that's the way of it for me.

TUESDAY 26 JANUARY

I chaired a session at the conference today. Strange to be doing it that way. Introducing speakers to apparently no one. Still, it went off OK. Then did an interview with French TV channel TF1. They're in Ireland to cover the lockdown.

WEDNESDAY 27 JANUARY

Devastated that the country got it wrong in December and we're now paying the price.

Someone texted someone in the lab to say they'd seen me on TF1 talking French. We had a look: I'd been overdubbed.

THURSDAY 28 JANUARY

Advocated strongly for antigen testing and also vitamin D supplements on Newstalk. Antigen testing may actually be a better predictor of someone being infectious than the PCR test. A positive has been shown to correlate with increased infectiousness, which is not the case with the PCR test. There is really good evidence correlating low vitamin D with more severe disease. People over 50 in Ireland are deficient in winter. So I pressed home that there should be a recommendation to take a supplement. It won't harm anyone, and at a minimum, it will help strengthen bones.

Finished the day with a lovely interview with Jennifer O'Connell as part of the Winter Nights *Irish Times* interviews. She asked me which I preferred, Oasis or Blur. That would be an ecumenical matter, I replied. She asked me would we be back in the beer gardens by June, and I said yes. I said it to give people something to look forward to, which is so important.

FRIDAY 29 JANUARY

Headline in *The Irish Times* this morning: Luke O'Neill says we'll be back in the beer gardens by June. Gulp!

New data from Novavax on their vaccine, which is another efficacious one. The results from the Johnson & Johnson trial also were released. Coming thick and fast now. This led to another interview on Newstalk on *The Hard Shoulder*. By then, the EMA had approved the AstraZeneca vaccine. Such a good message to give people ahead of the weekend: two more vaccines working very well and one more approved. And the weekend! Let's party! Er … maybe not just yet.

SUNDAY 31 JANUARY

Dull January comes to an end. Nothing but bad news on the case numbers. Tony Holohan said today that there had been more cases in the month of January than for all of 2020, with over 1,000 deaths and more than 100,000 cases. The price of Christmas. But the vaccine hope burned a bit brighter. Like the Pilot song 'January', I'm sick and tired of it hanging on me.

FEBRUARY 2021

MONDAY 1 FEBRUARY

I always like the first of February. In Ireland some say it's the first day of spring. It's also the feast day of that great pagan goddess St Bridget. I can see the hawberry buds on the trees in the front garden turning red.

There are now clear models appearing to predict what will happen as the vaccine takes effect in the older and vulnerable people, and it looks good. If the top four priority groups in the UK are vaccinated, the death rate is predicted to fall by 90 per cent. This is the magic we want. I was upbeat with Pat this morning. It will happen and the fear will dissipate.

Some of the data that generated the press releases last week were revealed. All the vaccines are now known to prevent severe disease and death by 100 per cent. There were 20,000 people on the Novavax trial. Over 43,000 on the Johnson & Johnson trial. One word: remarkable. (I didn't say 'gangbusters'!) Also mentioned though how Merck and Sanofi had failed in their attempt to make a vaccine, which is a surprise because they are big vaccine companies. Just shows you – it's never a slam-dunk.

Had to do some marking today. Three final-year essays. Important for the students. They were all excellent, so I gave them good marks. Perhaps I'm becoming a softie.

TUESDAY 2 FEBRUARY

Yet more excellent news on vaccines – the AstraZeneca vaccine is reported to decrease transmission of the virus by 67 per cent. Now, this is great if it holds up, because it means that people who are vaccinated will also not be spreading it as much, and so the virus begins to go away. Another step down the tunnel towards the light.

Cork tomorrow. Yes, Cork! I'm being interviewed in person on *The Today Show* with Maura and Dáithí. I'm heading out of Dublin! The excitement. I have a letter giving me permission to travel and I put Sam on as my assistant, which is allowed. He can carry my bag but also help with some of the research on what I will be talking about. He's excited too.

Watched the last two episodes of *The Terror*. Spoiler alert: Captain Crozier stays with the Inuit. Weird but happy, sitting in animal fur with the sun streaming around him. Maybe that's another parallel with the pandemic: that we will all learn to live with the virus.

WEDNESDAY 3 FEBRUARY

Call today with GSK on their Otilimab therapy. This blocks an inflammatory protein called GMCSF, which was discovered by an old friend from Melbourne, John Hamilton. It seems to help in the over-70s, reducing deaths by 20 per cent. It's blocking macrophages and neutrophils in the lungs from doing harm. The disease looks to be a bit different in the over-70s and may well be driven more by GMCSF, so it's well worth pursuing. It was so exciting to see the data when they revealed it. Clear effect. Could be important.

Over to Heuston for the train to Cork. Disaster struck! They announced that the train was delayed, not clear for how long. Something on the line. What? Cows? Sam and I waited a good two hours on cold seats in freezing Heuston and eventually went home because I had to do a call with Ian Robertson for the *A Lust for Life* podcast. Had the train left on time I would have made it to Cork to record it. We discussed the situation on COVID-19 and Ian covered the mental distress aspect. Huge number of questions. Ian and I are like a double-act – I said Laurel and Hardy for some reason. This can't be right, as we're both tall.

THURSDAY 4 FEBRUARY

It was a good call not to go to Cork, as I enjoyed last night and had a mini lie-in this morning. One of those nice ones, where you're not bothered by much. Up for Pat, though. The International Chamber of Commerce are saying that the global economy will lose $9.2 trillion if the developing world isn't vaccinated. This is because global supply chains are being compromised, and there is a risk of new variants emerging. Pregnancy is also a high risk – a three-fold higher risk of hospitalisation and a higher risk of pre-term birth. Reassured any pregnant women listening that the vaccine was safe for them.

SUNDAY 7 FEBRUARY

I was so bored today that I watched the rugby. Ireland versus Wales. What's happening to me? I'm becoming a sports fan. I did snooze on the couch though. People are surprised when I tell them I have only limited interest in sport. Each to their own. I mean, what is the point of chasing after a funny-shaped ball? Not that I'm knocking people who like it. Does it relieve boredom? Or tap into something primordial? I read somewhere that sport is basically safe warfare and that we are evolved to fight. That may be true but where does that leave me? In the trenches hiding, I guess.

Also ironed 11 shirts. A record for me and very satisfying. There's something so soothing about ironing. The heat and steam and nice smell off the cotton. And smoothing something out. Good Lord, I must be desperate. And facing six more months of this kind of life?

MONDAY 8 FEBRUARY

A report has claimed that dexamethasone has saved 500,000 people. A simple drug like that, which I worked on during my PhD. Wow! On with Pat and I made everyone envious by saying how the UK expects to vaccinate all the over-50s by 7 April. Sob – why not us?

Spent four hours making a fundraising film for Senda, the Boston-based company I'm on the board of. It's for their website and for fundraising. The sound man, who I knew from other shoots, was amazed at what the company was doing, making an orally active vaccine for COVID-19 and improving anti-cancer and Parkinson's therapies. It was a bit of kick to describe it all. Senda could well help make vaccines available orally, which would make a huge difference not just for rich countries but the developing world too. Dead cool.

On with Claire Byrne again, in a bubble like the ones used at the recent Flaming Lips gig. I went in one, Bernard O'Shea went in another, and Claire interviewed us. I said I was doing it to help my musician friends who are desperate to get back to gigging, and if this is one way to do it, then go for it! You never know. Some smart promoter might take it on

Felt great when I got home. I made a ham, cheese and pickle sandwich to go with a huge cup of tea. Such excitement. Got a funny text off Brian: *Fuck's sake, Luke. You've got a PhD in immunology and they stuck you in a fucking bubble?!*

TUESDAY 9 FEBRUARY

Gave a talk to all the CEIST (Catholic schools) principals and it was magical – 217 people tuned in. A lot of principals. Started with prayers and then some interesting pieces from students. One was about the painting *Nighthawks* by Edward Hopper. I love that painting too, and the student wrote so eloquently about it – the isolation and loneliness and mystery of it. Drew parallels with lockdown, obviously enough. Someone also read out a Carl Sagan quote, which really got me: 'For small creatures such as we the vastness is bearable only through love.' A lot of talk about how spring brings hope. I'd almost become a Catholic again! Such well-meaning people. Gave them a big update and said how schools will be safe places. Adopt the measures and don't worry.

WEDNESDAY 10 FEBRUARY

Interviewed this morning in the RDS at the Irish Tourism Industry Confederation. It's the big annual meeting and this year it's mainly remote. Met Paul Kelly, CEO of Fáilte Ireland. They do such a good job for our country, but there is such anxiety now, of course. He said the whole industry was in a huge state of anxiety and uncertainty. Gave them the update on how by summer life should be back a bit anyway. Paul said the ad campaign will again be hugely focused on the home market, which will of course depend on lifting travel restrictions in Ireland.

THURSDAY 11 FEBRUARY

Pat and I covered the Johnson & Johnson vaccine, which will be used to vaccinate one eighth of the world's population. Told everyone how there is less and less evidence that you can catch COVID-19 off surfaces. This might lead to a policy shift. A bit of dirt helps the immune system tell the difference between friend and foe. I hope we don't see an upsurge in allergies and asthma in the coming years.

FRIDAY 12 FEBRUARY

Today was absolutely brill, just one of those great days that come along now and again. All four parts of my working life got attention. Communication – gave a talk to the Dún Laoghaire Arthritis Group and recorded two more podcasts at Newstalk. Data – Hauke gave a lab talk with some interesting ideas. Students – recorded two more lectures. New medicines – call with the pharmaceutical company AbbVie about recent developments in our lab: new possible targets for anti-inflammatories..

TUESDAY 16 FEBRUARY

Gave a talk to a Gaelscoil in Leixlip. The teachers were so grateful. Something different for the students. Learned a new phrase: 'Tá tú ar mhute!'

WEDNESDAY 17 FEBRUARY

Session with the Irish Gerontological Society. They said they never had so many people joining a session before. Thousands. And thousands who couldn't get in. Yet again, such an appetite, and obviously that is the age group who are most concerned. It was a thrill to do it. I told them all how to keep their immune systems healthy, both to ward off infection and to get ready for the vaccine. Good to give practical advice as well as science.

THURSDAY 18 FEBRUARY

Worked from home today. It was tough. Too many damn tasks with other ones hanging over me. Still, got through them.

On with PJ Coogan on Cork FM. Looks like the world might be turning a corner, with case numbers everywhere falling. But it's a mess in the US, where the number of deaths now stands at 500,000. Getting close to the 1918 pandemic, which killed 670,000. A travesty.

Also covered the COVID-19 endgame, which he was delighted to hear! I said how it might be in sight.

Was on *Prime Time* with Miriam. She had read last Sunday's interview with me in the *Sunday Times*, where I mentioned how me and Tony Connelly were big pals and how we played music together in London and Cambridge. We used to drink and play in a pub called the Geldart. I spoke about how there was an Irish landlady called Mary there who used to feed us, and when I got the lectureship back in Trinity she was delighted for me. She said that that was everyone's dream – to go back home. I also mentioned how Andy Irvine (famously of Planxty) came to do a gig one night. We had a lock-in (as was normal) and at 1 a.m. there was a knock on the door. We thought it was the police, but it was Andy. He said he came back as he'd forgotten to say goodbye to Mary.

We chatted about the AstraZeneca vaccine, which some people won't take, because they say it's not as good as the others. I reminded everyone how it is 100 per cent effective at stopping severe disease and death, so take it! For God's sake, take it!

MONDAY 22 FEBRUARY

Pat and I discussed nicknames for the different strains: Eric, Nelly. Nelly is more transmissible. Eric might be somewhat resistant to the vaccines. Pat made a joke: what about antibodies to Nelly? Aunty Nelly? Someone emailed me after to thank me for making the variants easy to understand.

Israel reopened yesterday with restrictions, including needing a vaccine certificate to be allowed into a gym. Life is slowly returning there.

TUESDAY 23 FEBRUARY

Gave another mega talk on COVID-19, this time to Comfort Carers. This is a huge organisation caring for people in their homes. Loads of questions, about the vaccines mainly. Several from concerned pregnant women, so I was happy to reassure them.

Government announced that nothing will change until 5 April. Six more weeks of this ragged life.

WEDNESDAY 24 FEBRUARY

Up to Newstalk. Recorded an interview with Adrian Hill on the Oxford vaccine. It was great! He gave a really good account of it. How they realised early they would have to give it away at cost until the pandemic is over. How the mess-up in the trial was explained by the gap between the two shots – 12 weeks is best. How the approach they are using is good for booster shots with new variants. Can't wait for it to be broadcast.

THURSDAY 25 FEBRUARY

On the show this morning I said that the government should allow golf, tennis and group cycling. It's so important for people's mental health and overall well-being. They'll have to do something for people, otherwise there will be a revolution.

We moved onto the Valneva vaccine, which is made in France. It's the whole inactivated virus, so could well work against all variants, as it will recognise lots of parts of it, many of which won't have changed. The other vaccines work against only one part: the spike protein. If the spike protein changes too much the vaccine might fail because it will only train the immune system to recognise the previous version. This is less likely to be a problem with the Valneva vaccine as it will train the immune system to recognise lots of parts of the virus and they won't all change in variants that arise. They used a similar approach for their vaccine against Japanese encephalitis. The vaccine has an immune booster that goes through TLR9 – one of my favourite immune proteins that really gets the immune system going.

Also discussed how quarantine hotels weren't ideal, because spread happened from them. There was positive pressure on hotel rooms: open the door and the air rushes into the corridor, bringing virus with it. The virus actually jumped across the corridor into the room opposite, infecting the people in that room. I felt it was important to cover this, as we are about to start hotel quarantine here.

Saw on the news that the global death toll now exceeds 2.5 million.

FRIDAY 26 FEBRUARY

The interview I did with *Mature Living* magazine came out and caused a flurry. My Grafton Street assault was mentioned. *Irish Daily Mirror* and *Claire Byrne Live* wanted to know more.

SATURDAY 27 FEBRUARY

I found today's events disturbing, to say the least. A protest against lockdown on Grafton Street turned into a riot. Garda were assaulted. Someone got footage of a guy shooting a firework at the Guards. Appalling scenes. I hate violence of any kind.

SUNDAY 28 FEBRUARY

Well, now! I was on RTÉ Radio One with Tony Connelly for *Sunday with Miriam*. She interviewed us and we talked about how we met and how much music meant to us. People sent in photos of us playing from years ago. Claire O'Connell sent in a message saying how when she met Tony and said to him that she heard he was in my band, he had said, 'Oh no, Luke is in *my* band.' Big response to it. I guess people are surprised but there was also delight. I suggested they play '500 Miles' by The Proclaimers, as that was one we used to play together. I said how Tony's brother Stephen had put us on to them. He'd seen them on a music programme called *The Tube*. Stephen tragically died of cancer a few years back. Miriam dedicated the song to him, so a wee tear welled up! Good Lord, I think it's 20 years since he passed. All in all, a lovely thing. Yet another unusual thing in these strangest of times. '500 Miles' sounded great in my kitchen, and the memories rushed back at me.

Wrote a piece for *The Conversation* on how long protection might last against COVID-19. A dense sea fog then rolled in. Went for a walk and it was eerie. There was a white rainbow in the fog, a rare event.

Gen from *Claire Byrne Live* asked me to talk about being assaulted on Grafton Street. She also told me that Wayne Coyne, of the actual Flaming Lips, had seen the picture of me and Bernard in our bubbles and sent it out in Instagram. Like, wow!

MARCH 2021

MONDAY 1 MARCH

The year seems to be gathering pace. March already. St David's Day. Definite evidence of a bit more light.

Did a pre-record with Claire Byrne and mentioned the time I was accosted by an anti-vaxxer in Galway. She asked my view of the riot. People are frustrated and angry but that doesn't justify the violence, was my reply.

Then straight over to *The Tonight Show* on the same topic: Marg had five nasty messages left on her answerphone at work. On the Dart a little old lady berated a big guy who was shouting at me, and he backed off.

Feel a little uneasy revealing all this, but I hope it

does some good to call out these nasty people. I said how I knew of women immunologists in the UK who were so vilified and threatened with sexual violence on social media they closed their accounts. I said how we must keep at it – keep up the fight. Attacks are horrible, especially against people who are actually trying to help.

TUESDAY 2 MARCH

Still getting some interesting emails over the past few days – really nasty ones go straight to spam, so it's mainly nice ones:

> A dear friend is getting married in June. She's gone from joy to despair and every emotion in between. But every time you made a positive observation we'd text each other saying 'I love Luke'. At this stage you're liable to get an invitation to the wedding (numbers permitting!).
>
> ············
>
> I'm an eight-year-old girl. I thought you'd be a really cool person to Zoom with for my class. This is the first email I have ever sent.
>
> ············
>
> My mother is willing to escort you around Dublin. The woman has been known to turn lightning around. She'll also teach you to play poker. She'll

accept payment in kind: Cork Dry Gin, tonic water and slice of orange.

....

THURSDAY 4 MARCH

If you are carrying certain Neanderthal genes, they put you at higher risk of severe COVID-19. It was disputed for a long time whether we humans had sex with Neanderthals when we encountered them in Europe. They were after all brutish cavemen, or so it seems. Well, we did. And we still carry the genetic legacy of that. Some of the genes we inherited from them are involved in the inflammatory process. It looks like they might become over-active when we're infected with COVID-19. One of them makes a chemokine (an important immune protein that attracts white blood cells to site of infection to fight the germs), so that might be a new target for COVID-19 patients. Now there's something. A gene we've carried for thousands of years, that we picked up from sex with a caveman, hiding away in our genome, is now a target for SARS-CoV2, a virus that only infected us last year. You couldn't make it up.

Also discussed how the evidence grows and grows that COVID-19 is a disease passed on by super-spreaders. Ten per cent of cases are infecting 80 per cent of people. Why would that be? Some just talk more loudly, so more virus comes out. Speaking loudly has been shown to expel 50 times more virus than talking softly. And as for singing – that's been shown to release 99 times more virus than talking. This explains why choirs were sadly such dangerous environments.

FRIDAY 5 MARCH

Did an interview today with *The Irish Times* on my top songs in lockdown. I went for lots of uplifting ones:

'Mexico' by Mundy – well I have to have that one, right?

'Open Arms' by Elbow – Stevie! Sob!

'One Day Like This' by Elbow – '... would see me right'. Too right!

'Cape Cod Girls' by Baby Gramps – sea shanties have become popular, maybe because it feels like we're on an interminable voyage with COVID-19. I picked another: 'Rolling Sea' by Eliza Carthy. A sea shanty by a woman for once!

Two Beatles songs: 'Getting Better' because it can't get any worse and 'Here Comes the Sun', just because.

'Me and Magdalena' by The Monkees – because, unlike most of their other stuff, it's a lovely song.

'N17' by Tolü Makay – a great version that was done at the New Year's Eve show.

'A Hero's Death' by Fontaines D.C. – because life ain't always empty.

While I'm at it, my top five albums are *Sergeant Pepper*, *Abbey Road*, *Hounds of Love* by Kate Bush, *Balance and Control* by Scullion and *Sunshine on Leith* by The Proclaimers.

SUNDAY 7 MARCH

Another window visit with Desiree today. Sam came with me to see his granny. Bitterly cold again, so she kept worrying we were freezing outside. She wasn't in wonderful form today – didn't know my name. But yet again, a lovely smile as I left.

Went for a walk in Dún Laoghaire. I said to hell with it, Sam, let's make a day of it and get an ice-cream. Got one from a van near the People's Park. The man recognised me and insisted that the ice creams were on him. He said he'd seen me on the TV talking about the abuse I'd got. Yet more evidence of the kindness of people.

MONDAY 8 MARCH

Much brighter morning. Spring is here alright! Spoke about long COVID. It's been renamed 'post-acute sequalae of COVID-19' or 'PASC'. Docs love to name diseases. $1 billion is being invested in research. Also said how there is anecdotal evidence that vaccination is providing some relief for those with PASC.

WEDNESDAY 10 MARCH

Johnson & Johnson has been approved in the EU. Vaccine number four. Four vaccines! I still can't quite believe that.

THURSDAY 11 MARCH

The trials for vaccines against the variants will be much faster because it's the same technology that will be used. And we now know a lot more about why COVID-19 is more severe in men. Looks like testosterone is boosting the levels of ACE2 in the lungs and so men get more infected. In women, oestrogen seems to protect them from inflammation. Pat asked should we give men oestrogen and I said, well, that would have other effects too.

The US is now vaccinating over 1 million people a day. This is the biggest global vaccination campaign in history. Every country is involved. The goal is nothing short of vaccinating the entire Earth's population. I have to pinch myself when I think of the huge scientific success this is. Bring it on.

FRIDAY 12 MARCH

Got an unusual letter today that is worth recording:

Dear Professor O'Neill,

Many women hang on your every word and like Doctor Tony you have certainly helped thousands since this deadly virus began to stalk our land. I live with my ageing husband on our small farm in the kingdom. I am wondering if alternative medicine might have a role in combating COVID-19. I have studied aromatherapy and I believe tea tree oil might have some benefits?

My late mother also had a cure for any illness. At night I still follow her advice. I boil onions and cloves of garlic and spread this paste on my husband's chest and my own. In the morning I put a couple of slices of onion in my socks before I venture out for my constitutional. I also spray my body with vinegar, then put some soap up my nose. Of course I also wear a mask and a pair of plastic gloves. I take castor oil to keep myself regular. You should really try garlic and onions. They would certainly discourage young thugs from attacking you on Grafton Street.

Sadly, my husband still believes alcohol will cure everything. He insists on sprinkling brandy on his cornflakes in the morning, and I am convinced he is keeping his spirits up during his frequent visits to the hayshed.

Our son is faring slightly better. He believes in the curative powers of cannabis. As a 12-year-old he left a plastic statue of the Virgin Mary in his bedroom, which I came upon. Her crown had been covered with wire gauze and a plastic straw protruded like a pipe from her bottom. At the time I was very vexed but now he smokes it openly in the farmhouse and I am happy to inhale the fumes.

Like all of us, I am hoping for the best and say a decade of the rosary every evening for the country and our salvation. My son is worried that the vaccine will lead to us all being injected with Chinese microchips. Even worse, he tells anyone who will listen that Micheál Martin invented the virus in order to take credit for saving our lives and convince us to vote Fianna Fáil. I just don't know what to do. Should I stick with my tried-and-tested traditional remedies or put my faith in modern medicine? I have no one else to turn to and I hope to hear from you soon.

P.S. Have you ever considered a move to Hollywood? At the very least you should have your own chat show on RTÉ. You would be a big improvement on Tommy Tiernan.

Yours, etc.

I wonder if Tommy Tiernan wrote the letter?

SUNDAY 14 MARCH

Had a few cans last night, so was a little fragile this morning. Checked my phone on waking. I'd missed calls with Sky News, RTÉ news, Newstalk and Al Jazeera. What the hell? Checked the headlines. The AstraZeneca vaccine has been stopped in Ireland because of four cases of unusual blood clots in Norway. I got back under the duvet.

MONDAY 15 MARCH

Did a lot of reading about what's happening with AstraZeneca. These blood clots are extremely rare events – 1 in 160,000. You're more likely to be knocked down on your way to the vaccination centre than be harmed by this vaccine. On with Pat and laid into it – said it was a big mistake to stop the roll-out as the risk was so low compared to COVID-19, and people would now not be vaccinated. They might get infected and have a severe disease. What's being applied here is the cautionary principle, but this is all wrong. If you take an action out of a sense of caution, that action shouldn't risk causing harm in another way, which is exactly what is happening here. I also said it could increase vaccine hesitancy. It's a bump in the road. I think it's because of the intense spotlight on the whole vaccine campaign.

But the Novavax vaccine is giving great results in

trials. And a study on healthcare workers has shown that the Pfizer vaccine is preventing transmission by as much as 70 per cent. Yet again, that vaccine is top of the class.

WEDNESDAY 17 MARCH

A lazy Paddy's Day if ever there was one. Drank cans of Guinness in the front garden with Sam. Sun shining. Marg, Brian and Claire went for a swim in the Forty Foot while I held Marg's bag. They said it was refreshing. I said they were mad.

THURSDAY 18 MARCH

On *Morning Ireland* – the AstraZeneca pause is still the big news. I explained how the clots were so rare and not typical because of where they are, which is mainly in the brain and abdomen. The EMA were bound to say the vaccine should continue to be used, because of the very low risk. A survey had shown that 38 per cent of people in Ireland were now reluctant to take it. You see? I predict they will say there is a possible link to the vaccine. None of this is unusual when it comes to new medicines or vaccines. Give it to 30 million people and you're bound to see a few adverse events.

We talked about how the FA Cup will go ahead with fans. They are letting 10,000 into a 90,000-seat stadium. Whether they have to show evidence of vaccination is being discussed too. Fans cheering in Wembley again will be a sure sign we're on our way from misery to happiness.

Did an interview with RT, the English-language channel in Russia. It was for a programme called *Worlds Apart* and was a one-on-one interview with Oksana Boyko for almost 30 minutes. She said they had recently interviewed Richard Branson and Richard Dawkins. We covered all the key issues around COVID-19, particularly the EU's performance with the vaccines. She asked me about Sputnik and I said it was an excellent vaccine and could we have some please?

Then out to *Prime Time* at 7 p.m., where they showed a film by Yvonne Murray from China on the origin of the virus. This is still not certain. Yvonne filmed in the Wuhan market and at a press conference with Chinese officials. Not many answers. I said it was disconcerting that months after the first case we don't know exactly where the virus came from. I interviewed Adrian Hill – asked him what he thought. He said it was right to examine these rare cases but that there was no need to stop the vaccination campaign, which is what the EMA are saying too. So, our government has gone against the EMA and the MHRA in the UK. Brave of them.

A Norwegian scientist talked about what happened in the four people with clots. Germany is reporting cases of clotting too, so the numbers are rising somewhat.

FRIDAY 19 MARCH

Big meeting in the lab with Tristram, and also Roger, our clotting expert. He was on Zoom. It was dead good! The data are looking great. DMF is clearly blocking clotting factors both in macrophages and also in a mouse model of clotting in the lungs. Oooh … here's hoping. Might this be a whole new way to stop clotting during infection?

SATURDAY 20 MARCH

Well, I've made it. I was slagged off big time on *Callan's Kicks*: 'Luke O'Neill is everywhere.' He had me being elected Pope (*In the Name of the Pfizer*), winning an Oscar and also on the moon with Neil Armstrong, telling Neil he was outside his 5k. It was laugh-out-loud funny. Thanks for helping my immune system, Oliver. It's no wonder there's no far-right here – Oliver and his ilk are so good at taking the piss, it must destroy their credibility. Ah, they'll miss me when COVID-19 is gone!

SUNDAY 21 MARCH

Got a classic nasty email: *O'Neill, you are the most annoying man on TV. Your voice is grating. You've got what you deserved, being called out by Philip Nolan and Oliver Callan. Ha! Ha! I'm sick of your stupid smile and stupid phrases about beer gardens being opened. Just go away. From a very disappointed Trinity graduate.* Maybe I should ask him or her for a donation to our Trinity COVID-19 centre?

But on the plus side, I got a lovely letter from poet Gerald Dawe. He sent me a copy of his book *Looking Through You*, which is a quote from a Beatles song. He said how he loved my choice of 'Getting Better' at the end of the *Prime Time* edition. A nice counterbalance to the nasty stuff.

MONDAY 22 MARCH

Great session with Pat about the huge promise of the Sanofi vaccine, which is in development and looks promising. I told him it had a great immune booster in it – a biochemical called squalene. Can't beat getting the odd scientific term in! And how something we covered before is looking better. A drug called clofazimine, discovered in Trinity for leprosy all those years ago, can kill SARS-CoV2.

TUESDAY 23 MARCH

My old dad would be 100 today! Always think of him on his birthday. Big Joseph, as we used to call him, even though his name was Kevin. This was because of a song he used to sing from World War Two: '*This is the end of Big Joseph. Big Joseph is dying tonight.*' Cheery little number.

Chaired a big conference for the Royal Society today on cancer and the immune system. A welcome break from COVID-19. The biggest breakthrough in cancer is waking up the immune system to kill tumours. It's working and, dare I say it, it's getting better all the time. Some excellent speakers. And there was talk of using the new vaccine technology to beat cancer too. It's happening, I tell you, it's happening!

THURSDAY 25 MARCH

Bit of a red-letter day. I was presented with the George Sigerson Award by the UCD Biological Society. It's given to someone who has inspired others in the biological sciences, and I was truly honoured to receive it. Nobel Prize-winner Paul Nurse is a former winner. I gave a Zoom to a load of students and I told them my own history. Also gave them the big message from the movie *Little Miss Sunshine* – 'Do what you love and fuck the rest.' It went off very well and I'm beaming!

SATURDAY 27 MARCH

Spent the whole day at the virtual Irish Science Teachers' Association annual meeting. There were all kinds of interesting talks and presentations. Yet again I'm blown away by the commitment of our science teachers. There was a talk on how to handle difficult people. The speaker called them CAVE people (Completely Against Virtually Everything). I know lots of people like that. Message was to avoid them and try not to be one yourself!

I gave a talk too – usual update on COVID-19 to help the teachers as they get asked a lot of questions, from both pupils and parents. They are doing such a good job at keeping the schools open and looking after their pupils.

MONDAY 29 MARCH

Big vaccine chat with Pat. So positive (apart from the AstraZeneca debacle). We did a lot on the vaccine supply. So many production plants are now running at full speed all over the world, churning out these magnificent vaccines. All the companies helping each other: Merck helping Johnson & Johnson, GSK helping Sanofi. The list goes on. They're even making a nasal form of the AstraZeneca vaccine. Israel just looks better and better. Here we go!

Claire Byrne Live tonight was brill. Just keep going, keep going. They made models of the three variants (UK, South Africa and Brazil). I could talk about the science of them. All three have the Nelly mutation and so are more transmissible. They spread more and your lungs might take in more, making you sicker. The vaccines work against the UK variant. They work less well against the others but should protect you from severe disease. The other two have the mutation called 'Eek', which leads to antibodies binding less effectively. We're not fully sure so, as ever, we must be cautious. Public health measures will stop them spreading. We don't want any more variants, if you please. The virus only has so many tricks up its sleeve. It may have played its best shot.

TUESDAY 30 MARCH

Meeting on Zoom with the governor of Mountjoy prison. He has asked me to give a talk to all the prisoners and staff, which will be beamed into all the prisons in Ireland. Prisoners have done well so far, but they were hearing all kinds of things from their families and there is a level of vaccine hesitancy. I will reassure them all. Imagine being stuck in a prison during a pandemic with your loved ones outside. The Irish prisons have done a tremendous job at keeping infections under control.

WEDNESDAY 31 MARCH

A momentous announcement from the Taoiseach. Vaccinated people can meet up! The start of the vaccine bonus with a lot more bonuses to come. And people in nursing homes can have two visitors inside per week. Get ready, Desiree! We're coming out of it now for sure. I can really feel it.

Got home exhausted and slumped in front of the TV. Luxury! Watched a documentary about Jack Charlton. Ah, those glorious days. I was in Cambridge when David O'Leary scored the famous penalty against Romania. Jack passed last year and sadly had dementia. A heartbreaking scene had him watching the famous penalty shoot-out but he had no memory of it. He did, however, call out one name: 'There's Paul McGrath!' They had such a special relationship, and Paul said how Jack was like a father to him. Very moving. Yet again, I wonder will our drug with Roche impact on dementia and Alzheimer's.

March comes to an end. Apart from the AstraZeneca snafu, a great month. Vaccines are working, with more to come. Case numbers and numbers of people in the ICU falling too. And Israel looks promising.

Come on. We can do this.

APRIL 2021

THURSDAY 1 APRIL

I can't help but feel this new month will bring a constant stream of good news, with a few glitches here and there of course. Another amazing occurrence today. The US embassy asked me to give a talk to their staff. Word got out and finally staff from 27 embassies in Dublin attended, including six ambassadors. I mean, for heaven's sake! Gave them the Full Monty of an update. The Chinese embassy was also on, but I didn't talk about origins.

Then another talk for the Cystic Fibrosis Society of Ireland – an update on COVID-19 but also on how our anti-inflammatory drug might work in CF. Gerry and Oliver McElvaney got great data on that. They are the key CF doctors in Ireland, and it's been a productive collaboration.

SATURDAY 3 APRIL

Left it a bit late going to the butchers. Asked Noel if he had any lamb left. He laughed and said, would you ever go on with yourself? And then he sold me a turkey. Knocked €20 off the price. I came home with a turkey for Easter Sunday dinner. Marg was not impressed, but we got four dinners out of it. Cold turkey had me on the run, as John Lennon would say

SUNDAY 4 APRIL

Another Easter Sunday with COVID-19. Surely the last one! We never ever thought we'd be in this position a year ago. Went to see Desiree again. She had a lovely Easter hat on – they look after her so well there. One of the care assistants said he had done a PhD in immunology – worked on transplantation. We had a great old chat. Very well informed, as you might expect. An immunologist minding Desiree! Sure, how could it be better?

FaceTime call with Stevie and my sister, Hellie. He is with her in Brighton for Easter, which is brilliant. Two O'Neills cut off from us, but at least they can spend time together. Hellie is so loving to my two boys and a great influence on them.

TUESDAY 6 APRIL

Had to correct a lot of senior sophister immunology

projects today. The day job. They all managed to somehow mention COVID-19. Extra marks all round!

THURSDAY 8 APRIL

So, the EMA has officially said the AstraZeneca vaccine is safe. They said it is 'possibly' causing a rare type of clotting in the brain, but that the benefits far outweigh the risks. This is good. Tried to reassure listeners today. There will be bumps from time to time …

The UK and Canada are using up all their vaccines. They aren't keeping any in reserve for the second shot. This is because one shot has been shown to protect by a lot – 76 per cent for AstraZeneca and 80 per cent for Pfizer and Moderna. Kingston agrees that we should get it all out as soon as we can. I made the point clearly but will anyone listen? This is the quickest way to protect our people.

Great discussion today with the GSK immunology advisory group. They are doing a great job on their programme against COVID-19, working with lots of other companies to help optimise their vaccines. And working hard on new treatments. Such hope when you see a big pharma doing that. And they're all at it.

SATURDAY 10 APRIL

A sad day. Jane didn't win the election for new provost of Trinity! Linda Doyle won. Jane was eliminated after the first count. I accompanied her through it all. Not clear what happened. She is such an outstanding academic. Maybe that was the problem. I said she was too good for them. We opened the prosecco anyway. I will of course support our new provost, as Linda is great too. History was made: the first woman in the history of the college. And not only that – Rachael Blackmore won the Grand National, another first.

Got an email from a lawyer asking me to be an expert witness in a case of a woman being held in quarantine in a hotel despite been vaccinated. I wasn't needed in the end, because they let her out. The mandatory hotel quarantine seems to be annoying everyone.

MONDAY 12 APRIL

Big session with Pat on the clots – an autoimmune reaction to platelets in the blood. This is what causes the clotting. At least we know, and it can be treated with something called IVIG. They're calling it VITT, or 'vaccine-induced immune thrombotic thrombocytopenia'. Another new disease, but so rare. We also talked about the Valneva vaccine, which has given spectacular results in a Phase 2 trial. And how a drug called baricitinib, which blocks inflammatory cells, reduced deaths by 38 per cent on a trial. The good news keeps on coming.

But then in the afternoon there was another speed bump. The Johnson & Johnson vaccine was paused in the US because six people have had the same kinds of clots with what were seen with the AstraZeneca vaccine. Again, extremely rare. More likely to be struck by lightning. I think this pausing is happening because of the intense spotlight on the vaccines and also the regulators, and because it can be a lethal side effect. The National Immunisation Advisory Committee also have recommended that the AstraZeneca vaccine should only be used in the over-60s because young people are at such a low risk from COVID-19. This is causing consternation.

TUESDAY 13 APRIL

Huge conference on innate immunity with participants from 36 countries. It was supposed to be in Killarney. Bah! Look at all the money that would have brought in. It's organised by my former student Kate Fitzgerald, who is in the University of Massachusetts. I gave Tristram's data on clotting. A lot of interest, which is excellent for Tristram, as he might get a really good publication out of it. I will press for that.

THURSDAY 15 APRIL

Leo Varadkar was on with Pat just before me. Said he would listen to what I had to say about getting all the vaccine supply out. It must be done, I reiterated. Please Leo, please!

Another online conference – this one was meant to be in Boston. All about sex differences in the immune system. This is so important. Women are a much higher risk of autoimmune diseases, while men are at a higher risk of autoinflammatory diseases. Akiko Iwasaki gave a fascinating talk on why men do less well with COVID-19. She is leading the way on COVID-19 immunology in the US, and it was a privilege to hear her speak. We all felt it was such a shame that we couldn't get together in person.

In spite of all the ups and downs and ins and out, today is actually a good day. Over 1 million people have now been vaccinated in Ireland. The number of people in hospital and in the ICU is the lowest it has been since mid-December and will continue to go only in one direction. Philip Nolan said it was all very positive. And Pfizer have said we can have 500,000 doses of their vaccine now. We are on our way. Over 870 million people are now vaccinated in the world, in 154 countries. And Israel reported today that it has exited COVID-19. The vaccine has beaten it there. Exited COVID-19! Now there's a phrase, and it was coined by Eran Segal, one of Israel's leading immunologists. And the UK numbers look positive too – over 90 per cent drop in deaths and hospitalisations.

And yet, there have been over 140 million cases and as of today 2,947,644 lives lost to this virus. The virus that jumped from a bat into a human about 18 months ago. I still find it staggering.

But now I can see the light. There may be the odd blips ahead, and who knows what they will be, but we will get there. It's definitely beatable. And I need a holiday. Fifteen months of hard labour ... and no doubt people need a holiday from me.

Got a brilliant email today:

I work with a group of maintenance lads in Irish Rail. We've been following your Newstalk interviews and even reading your book. The normal read in our van at lunchtime is an English tabloid, so you've been definitely influencing us ... and bringing a little science to our lives. I never thought I'd see working men in their fifties reading science books, many of whom were only educated to primary level, so all very positive, Luke. The new catchphrase in the van is 'What does the data say?' – to much amusement! Thanks a million, Luke, keep up the great work. You are welcome to a cup

of tea and a lunchtime chat in our van when we're
all vaccinated.

....

And that sums it all up for me. When we're all
vaccinated indeed. Right, I must crack on as per. I
wonder what will happen next?

EPILOGUE

As I write this in August 2021, 83 per cent of our adult population is fully vaccinated and that will rise to around 90 per cent. Ireland's vaccine rollout has been a huge success story, ranked among the best in the world. We can all be very proud of both the logistics and the fact we put our trust in science as the main way to beat COVID-19. Globally, over four billion people have had a vaccine, more than half of the world's population, and all in less than nine months. These are very impressive figures, and they hopefully herald a return to the lives we all want to live.

On 30 July 2021, I managed to play some music live with The Metabollix. We hadn't played together in the one room since March 2020. We were well spaced out (physically speaking), and the performance was recorded for a special video to be sent to the alumni of Trinity College Dublin, who would normally be invited to spend a weekend, usually in August, in their old college. COVID-19 had yet again put paid to that, but Trinity wanted to maintain the connection with their graduates, so they filmed various performances at different locations on the campus.

We were filmed in Trinity's iconic Exam Hall, where there was plenty of room and a lot of ventilation. Looking down on us were paintings of notable figures from years gone by, including one of Elizabeth I, who had founded Trinity in 1592. I'm not sure what she made of the rock music we blasted out.

I felt like Rip Van Winkle. Rip features in a short story by American writer Washington Irving. He is a Dutch-American villager in a colony in the Catskill Mountains. He meets mysterious Dutchmen (are there any other kind?) and drinks some liquor with them. He falls asleep and wakes up 20 years later to a very changed world, having missed the American Revolution.

We are all Rip Van Winkles now. Slowly waking up from what feels like a long sleep. We have been through the worst pandemic in over 100 years, if not ever, that wreaked havoc all over the world. However, the deployment of the most powerful vaccines ever invented means that we have SARS-CoV-2 on the run. COVID-19 is now a preventable disease because of vaccination. But that didn't always seem certain, and, as you will have read, I was hugely relieved last November when the first – that made by Pfizer/BioNTech – of several vaccines showed remarkable efficacy in the clinical trial that tested it.

But there is more to be done. We need to get more vaccines to poorer countries not only to reduce the numbers dying from COVID-19 but also to stop variants of the virus emerging, which is our remaining major scientific concern. Variants have arisen that can break through the vaccines but, so far, not to the extent of causing severe disease and death. Booster shots are likely to be needed to maintain immunity, and those boosters may well involve variant-specific vaccines, a bit like the flu vaccine, which changes every year. Because of vaccines and better hospital treatments, COVID-19 will become a largely manageable disease.

From all the science that has been done, we know the rules of the game when it comes to COVID-19, and we can continue to use that knowledge in the ongoing fight, which isn't quite over yet. Like Rip Van Winkle, we must get used to a changed world. Many of us have suffered, physically and mentally, and we need to look out for each other now more than ever.

I am so proud of all that science has achieved and

hope we learn from this for the future challenges that we will no doubt face, be it another pandemic or that hugely important issue: climate change.

We have come through one of the most dramatic periods in living memory, and through our scientific ingenuity and care for one another we will prevail.

Acknowledgements

Thanks to Sarah Liddy for suggesting I produce a book from my diary, and Aoibheann Molumby and Djinn Von Noorden for making excellent editorial suggestions. Thanks to Andy Gearing, Brian McManus, Manus Rogan, Stevie O'Neill and Margaret Worrall for checking the text for me and recommending cuts or additions.